D1413130

THE SOCIAL INNOVATION IMPERATIVE

*Create Winning Products, Services,
and Programs That Solve Society's
Most Pressing Challenges*

SANDRA M. BATES

New York Chicago San Francisco Lisbon London Madrid Mexico City
Milan New Delhi San Juan Seoul Singapore Sydney Toronto

1 2 3 4 5 6 7 8 9 10 DOC/DOC 1 9 8 7 6 5 4 3 2 1

ISBN 978-0-07-175499-6

MHID 0-07-175499-7

e-ISBN 978-0-07-176015-7

e-MHID 0-09-176015-6

Library of Congress Cataloging-in-Publication Data

Bates, Sandra M.
  The social innovation imperative : create winning products, services, and programs that solve society's most pressing challenges / Sandra M. Bates.
    p. cm.
  Includes index.
  ISBN-13: 978-0-07-175499-6 (hardback : acid-free paper)
  ISBN-10: 0-07-175499-7 (hardback : acid-free paper) 1. Branding (Marketing) 2. Advertising— Brand name products. 3. Reputation. 4. Leadership. 5. Communication. I. Title.
  HF5415.1255.B38 2012
  658.8'27—dc23   2011037913

For my three amazing sons Joshua, Matthew, and Seth—and my bonus daughter Lindsey—the ultimate blessings and joys of my life.

CONTENTS

Chapter | 9   What Citizens Want   210

                Notes   226
                Index   234

FOREWORD

It's certainly exciting to see all the attention being paid to social innovation over the past decade. Nothing could be more worthy of our time,
resources, and focus than creating solutions to alleviate some of the
world's most pressing social problems. We are inspired by social innovation successes such as Grameen in microfinance, Teach for America in
education, and MinuteClinic in health care. And because we are not
simply dealing with the corporate bottom line, the rewards of success and
the consequences of failure are immense.

Social innovation—as with any type of innovation—is fraught with
challenges. No doubt the challenges of addressing the needs of the poor,
uninsured, and homeless are exacerbated in comparison to other types of
innovation. Yet it would be a mistake to conclude that social innovation
should not and cannot benefit from a systematic approach adapted from
the best practices applied to innovation in other domains.

At its core, social innovation is about satisfying a particular type of
need of a particular group of people. It is *not* about whether the solution
is created by a for-profit or not-for-profit entity. Social innovation shares a
requirement with other types of innovation—to understand the needs of
those affected, to devise innovative solutions to address those needs, and
to deliver the solution so that its value can be realized. As such, it can also
share in the insights gleaned from innovation in other spheres. The truth
of this statement can be seen in a recent personal experience.

My wife and I felt fortunate recently to discover an innovative company in the health-care space. We had visited our pediatrician on more than one occasion to get a diagnosis and treatment for an unknown growth on the finger of our youngest daughter, Julia. The diagnosis: "Don't know." The treatment plan: "Wait and see." We had waited—hence, multiple visits. It was now time to escalate the medical care to a specialist.

We are not uninsured, but we may as well have been. Our search for dermatologists in our insurance network indicated that the closest dermatologist was more than an hour's drive away and that she would be happy to see us at her next available appointment—more than six weeks out. The search for an out-of-network dermatologist was no more reassuring. It would also take many weeks to be seen and would cost more than $150 for the visit. Ouch!

So we took a small but calculated risk on a company called JustAnswer.com. JustAnswer provides an online platform for people to ask questions of experts in a variety of professional fields, including health care. Its stated mission: "equal access to Experts for all" via a platform that provides "a fast, affordable and convenient way to connect with Experts." Sounds like social innovation to me.

So we asked our question, shared information and images with our online expert, Dr. Nair, and got our answer. A common wart. We took Dr. Nair's advice, and the wart was gone within a few weeks. With Just-Answer, we were able to engage Dr. Nair at our convenience—not his since it was after midnight for him. We were able to "see" Dr. Nair immediately and were able to obtain our diagnosis within a couple of hours. And the cost for getting our answer? $19. We happily provided a tip for Dr. Nair.

Whether JustAnswer is a for-profit or not-for-profit entity is beside the point. The point is that—at least for us—JustAnswer provided more affordable and convenient access to health-care services than was available to us with the standard health-care system. And the point is also this: Despite the focus on affordable health-care access—a social challenge—our underserved needs could have been readily gleaned from a simple conversation with us about our frustrations with current health-care options for the problem we were facing. With these needs in mind, JustAnswer has devised and delivered a classic "good enough" disruptive innovation. In other words, JustAnswer has applied the best practices of innovation in other spheres to innovation to address a social problem. And you can too!

This is the value that Sandra Bates provides in *The Social Innovation Imperative*, which offers a practical and systematic innovation process that builds on innovation's best practices while still recognizing the unique challenges facing social innovation initiatives. Over the past several years, I have had the pleasure of working alongside Sandy as we led innovation projects and trained innovation champions in companies from a diverse variety of industries. I know the expertise, experience, and passion she brings to social innovation.

Social innovation does not need to be a mysterious "black box" process. Like other types of innovation, it can be approached in a systematic manner, and Sandy shows how in this book. Sandy is not simply laying out best practices that she hopes will work. Rather, as she shows through countless illustrations, the methodology, tools, and templates of *The Social Innovation Imperative* have been tried, tested, and refined through years of application.

In our training work together, one comment that we received often was how refreshing it was to get practical guidance and tools concerning how to actually *do* the work of innovation. Though a lot of this praise was rightly the result of the innovation process we were using, it was also in no small part a response to Sandy's desire and skill to make the theoretical practical through hands-on tools and templates. As she has done with innovation training, Sandy does in *The Social Innovation Imperative*. She makes the theoretical practical, and the reader is sure to benefit from her extensive and practical experience.

Peter Drucker once wrote that, "Above all, innovation is work rather than genius. It requires knowledge. It often requires ingenuity. And requires focus."[1] If you are wondering, "How can I work smarter to focus my ingenuity to deliver innovative solutions to society's most pressing social problems?," then this is the book for you. Let the work begin.

Lance A. Bettencourt, Ph.D.
Author, *Service Innovation: How to Go from Customer Needs to Breakthrough Services*

ACKNOWLEDGMENTS

This book is the result of ideas that have been forming for the past 10 years of working with some of the best and brightest innovators in the corporate world. After seeing success after success in applying the methodologies of Strategyn, Innosight, and others, I felt an obligation to apply them to bigger issues—the social issues that plague our country. I am grateful to Knox Huston, Daina Penikas and the entire team at McGraw-Hill who saw the possibilities that this book could bring to social innovation. Special thanks goes out to Mark Sowers for his insights and tireless help in working through the details of the book and to Francesca Forrest for her amazing editing skill.

I would like to thank my brilliant colleagues at Strategyn with whom I've grown, learned, and delivered the work we all love over the last several years. I am grateful to Bob Pennisi for always being there and supporting my crazy ideas, Lance Bettencourt for continuing to challenge me and help me to see things in new ways, and the pure genius of Eric Eskey. These men demonstrate the notion of integrity in work and life. They truly have the gift of innovative genius and generous hearts that share willingly with those around them. I was also blessed to work with some of the finest innovators from the corporate world who became fellow innovation zealots including Jeff Baker, Scott Burleson, Brandon Knicely, Rick Norman, Mike Lee, and Zac Lyon.

I was also very fortunate to work with our international colleagues. I respect them greatly for the challenges they overcome in order to bring a solid methodology for innovation to their countries. These colleagues include Chris Lawer and his amazing team in the United Kingdom, Bruce Burton and Sue Moorhen, Petr Salz, Kuba Karlinski, Bruno Levy, Maurizio Beltrami, and Martin Paterra. I hope to continue working with these teams to bring the Social Impact Framework beyond the borders of the United States.

I'd also like to thank the founder and CEO of Strategyn, Tony Ulwick, for believing in me, mentoring me for nearly a decade, and demonstrating the discipline it takes to make great ideas become a reality. He is a gifted, talented innovator and I am grateful for the time I spent being part of the Outcome-Driven Innovation® movement. It was a wild and crazy ride, and I wouldn't trade it for anything.

One of the strangest acknowledgments that I really must make is to a man I've never met—Seth Godin, the prolific author of *Linchpin* and *Poke the Box*—who introduced me to the "lizard brain" and taught me to squash it so that I could finish this book. I thank you Mr. Godin for sharing your gift and your art with the world.

I also would like to thank the many clients and colleagues who have been on this journey of exploration of social innovation: Ellen Domb (TRIZ expert extraordinaire), Joe Grieshop from Knovation, Marion McGowan of Lancaster General Health, Jeff Hynds with Ingersoll Rand, Margaret Laws of the California Health Care Foundation, and John Hall at CABA. This group was instrumental in helping me work through the application of the methodology.

There are several corporate clients who gave me great insight and support as we explored advances in the innovation process: the team at Microsoft including Dave Wascha, Mark Smolinski, and Bobbie Bakshi, David West and Amy Neumann (Rockwell Collins), Doreen Tho (Hospira, Abbott Labs), Thom Nealssohn at Masco, Michael Reynolds (WellPoint, Cigna), and Sarah Caldicott-Miller (author of *Innovate Like Edison*).

I am also grateful for the mentors who supported me and from whom I learned about innovation, marketing strategy, product development, and so much more: my honorary brother, and long-time mentor, Andrew Johnson who has been a continual source of wisdom and support; Steve Lindstrom, my first executive mentor who taught me to be fearless and relentless in pursuing that which can make a difference; and Terry Richey, the man who introduced me to the fun and exciting world of ideation. These men have had a significant and lasting impact on me personally and on my career.

My deepest gratitude goes to my family who provides a never-ending supply of support and encouragement: my husband Rob who is my anchor, and the source of my sanity and strength; my amazing boys, who became young men when I wasn't looking, and my lovely bonus daughter who has blessed our family with genuine joy and happiness. I am proud of all of you, and I thank you for believing in me and supporting my work all these years. I am fortunate to have both parents (Bob and Irene DeSanto) and in-laws (Jack and Margaret Bates) who are very much a part of my life and who have supported my efforts even when they didn't really know exactly what I did for a living (hopefully this book will help). And then there are my two terrific sisters (Sarah Salter and Dr. Gina DeSanto) who are a constant source of laughter and delight, such a joyful part of my life. I also thank my brother-in-law, recently retired Colonel Bill Salter whose work has been an inspiration for me in applying these techinques to the military.

Finally, I am grateful to God, my guiding force, for blessing me with everything that brought me to this point in my journey—the skills, the education, the experiences, and the amazing people I've encountered along the way (teachers, mentors, clients, students, and friends). My goal is to use these gifts to create good in the world; a goal for which this book is a first step.

INTRODUCTION

Meet Boyo, a child from a rural part of South Sudan near the village of Doro. He lives with his mother and five other children in this desolate part of the country where hunger is a way of life. Famine and insufficient drinking water are felt by all the people of this area, but none so much as mothers of babies and young children. These mothers cannot produce enough milk for their young children and babies, and with the lack of clean water, they cannot even use the powdered milk that is provided by the food aid program. There is little hope and much sorrow for a mother who can do little to alleviate the suffering of her precious children. In this part of the world, when a child gets as severely malnourished as Boyo, a slow and certain death usually follows. Boyo's mother has seen the results of this condition among other families in the village. The only option is to seek treatment at the clinic which is a two-day journey. If she takes the journey, she will have to take the chance that Boyo will be strong enough to make it. She also worries about her other children who must be left with a friend while she and Boyo are gone for an undetermined amount of time. But she knows she must go before he gets worse.

At the end of the long journey, they approach the compound where the medical clinic is located along with a new building called the Village of Hope Nutrition Center. As they approach the area, Boyo and his mother are greeted warmly by the missionary nurses from the center. Village of Hope was a project funded by SIM International and has been a beacon of hope for the people in and around Doro. It is staffed by a team of missionaries, nurses, and doctors who provide medical services, AIDS awareness and education, as well as nutrition education and treatment for malnutrition. One of the key resources available at the nutrition center is the miraculous treatment for malnutrition called Plumpy'nut. This breakthrough in food has become a major force in battling malnutrition around the world. Plumpy'nut is a high-calorie food consisting of a nutrient-fortified, peanut-butter–based paste that comes ready to eat in a small foil packet about the size of a deck of cards. Plumpy'nut requires no water preparation or refrigeration and has a two-year shelf life, making it an ideal solution to malnutrition in the difficult environmental conditions where it flourishes. Invented in 1997 by Nutriset, this innovation is widely acclaimed to be as significant to the fight against malnutrition as penicillin was in combatting illness and preventing the spread of disease.

Before Plumpy'nut, treating malnourished children in rural areas required a visit to the medical clinic/hospital where the child was fed intravenously for several days, utilizing a much needed hospital bed in what was generally an overcrowded medical facility, and keeping the child's mother from being able to care for her other children.

Undoubtedly, Plumpy'nut saved Boyo's life and has had a similar impact on thousands of children throughout the world. In Figure I-1, just one of these children, with his mother, is shown outside the clinic in Doro after receiving his Plumpy'nut; children really enjoy the taste of this food, making Plumpy'nut even more effective in overcoming the malnutrition of the child.

Given the dramatic, positive impact that innovations such as Plumpy'nut can have, it is imperative that this kind of breakthrough innovation become a replicated process in the social space to tackle other social issues. It is imperative that we create the processes, platforms, and programs that will spur this kind of thinking and creativity on a global scale with the same energy, vigor, and determination that goes into inventing commercial solutions such as the next iPad or next car.

The Social Innovation Imperative offers a refreshing new framework that incorporates state-of-the-art innovation theories and techniques with new components, strategic models, and the latest thinking on social innovation

Figure I-1   A Family Seeking Help at Nutrition Village in Doro

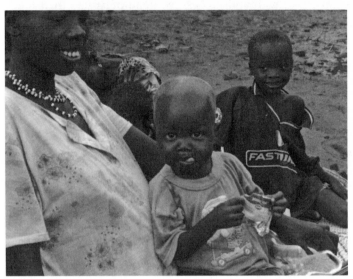

to produce a methodology designed for the unique challenge of solving major social issues. My goal is for this book to become a valuable resource on how to approach innovation in the social sector—whether for health care, aid for disaster victims, education, wellness, government effectiveness (including economic development), natural resource preservation, or any other vital issue. My hope is that nonprofits, Non-Government Organizations (NGOs), foundations, corporations that are seeking to generate social value, and government agencies can put the framework to use today for the good of humanity.

This has been my personal mission for the past several years. After spending the last decade working with major corporations and helping them create new markets for technology, consumer goods, and services, I felt that this same level of success has to be applied to the social space in the hope that we will soon see the type of innovative launches in the social space that are a dime a dozen in the commercial space. It is the only way to ensure that people around the world get to live the kind of life they should be living, one of personal growth and well-being, dignity and prosperity.

During the past several decades, having realized that they could not "cost-cut their way" to growth and success, companies turned their focus to innovation as the way to organic growth. As a result, the field of innovation has exploded, and there are now dozens of methodologies that are highly effective in addressing different aspects of the innovation process. There are three key parts of innovation—identifying the unmet needs of the people (investigation); creating clever, breakthrough solutions to meet those needs (ideation); and making those solutions happen (implementation). Some of the more successful methodologies focus on identifying unmet customer needs, which companies can then target with products and services. In fact, David A. Garvin from the Harvard Business School has stated that, "Studies comparing successful and unsuccessful innovation have found that the primary discriminator was the *degree to which user needs were fully understood.*"[1]

In the course of helping dozens of organizations create new products and services, I've become more and more convinced that the same tools and techniques that are used to meet customer needs *can*, and should, be used to address the social challenges plaguing our world generation after generation. After all, if a new laptop launch is delayed, the company's market share may drop, but people don't die. Innovations that help provide aid to disaster victims, on the other hand, can mean the difference between life and death. Innovations like Plumpy'nut have saved millions of children's lives in the most impoverished parts of the world. Innovations in our education system could mean the difference between a life of success and a life that never achieves its full potential. These scenarios are where innovation techniques can have the most impact, yet less than a third of the country's innovation resources are spent on these issues. It is imperative that significant energy be exerted on innovation that is focused on solutions that will make society stronger, make our planet healthier, help our children have a better future, and help those most in need.

Why don't we do this as a matter of course? Bill Gates stated in a *Time* magazine article in 2010, "while Capitalism has improved the lives of literally billions of people around the world, there are still 'a great many that do not benefit because they have *needs that are not expressed in a way that matters to markets* (emphasis added)."[2]

So, why haven't social innovations like Plumpy'nut exploded in recent years? In many cases, as Gates stated, it is because the needs are not understood by the corporate forces that could make a difference, and the connection between the needs of the people and the needs of the corporate entity has yet to be made.

Imagine innovation programs that target hunger alleviation, disaster relief, education health-care delivery, or conserving our natural resources. Imagine the impact the leaders of our government could have if they took a systematic approach to understanding their citizens' true needs and then used this information to work with government agencies, corporations, and nonprofits to meet those needs. Imagine the impact of disruptive innovations that would enable students to learn more effectively, regardless of where they live or what kind of school they go to. Imagine health-care innovations that do away with the need for nursing homes for seniors; innovations that enable seniors to live their entire lives as productive independent citizens. Imagine a world where resource conservation is the norm and that the physical waste generated by every company and household is reduced to zero.

This type of revolution in innovation in the social sector is well within reach. The techniques are available now. They have been proven in corporations over the last two decades and have resulted in substantial improvements in the success rate of innovation programs—some claim to achieve as high as an 86 percent success rate. *This process and "needs" information can in fact become a common language, a point of agreement around which people from different sectors can rally to create solutions that will have a major impact.*

Defining Social Innovation

Social innovation is the process of addressing the world's most pressing challenges with "novel solutions . . . that [are] better than current solutions, new to the world, and *[benefit] society as a whole* and not just a single entity."[3]

Because social problems are so intractable, the process for addressing them must be driven by the collaboration of people across different industries, disciplines, and regions. Andrew Wolk, CEO of Root Cause, a consultancy for nonprofits, points out that, "Although annually more than a trillion dollars are spent on millions of American nonprofit and government institutions . . . there is still not significant progress on social issues in the U.S."[4] He says that the only way to bring about real change is to *create an innovation and collaboration platform that draws together government, business, and nonprofits.* This sentiment is echoed among those working on innovations in our education system, those working to create economic growth, and those working on conserving our environment. It will take the joint efforts of the corporate sector, nonprofits, government agencies, and regular citizens to create solutions to these massive issues, a notion called Collective Impact. The vision is that social innovation becomes *everyone's* responsibility.

This notion has reached the halls of Harvard and the *Harvard Business Review*, where renowned business strategists Michael Porter and Mark Kramer provide a harsh assessment of the business world's current involvement with social issues: "Business increasingly has been viewed as a *major cause* of social, environmental and economic problems. *Companies are widely perceived to be prospering at the expense of the broader community* (emphasis added)."[5] The article details the negative impact that the current unsustainable, short-term-oriented business approach has on society. The authors define "shared value" as the value created by organizations that benefits both the organizations and society as a whole and that settling for anything less than shared value *underutilizes* the power of capitalism. Social innovation is no longer something that is merely "nice to have." Now it is a *business mandate*, with promising benefits if it is embraced. The authors note that, "The connection between societal and economic progress has the *power to unleash the next wave of global growth.*"[6] In other words, not only is it worthwhile to pursue social innovation for altruistic reasons, but it is now economically advantageous for corporations to do so. Such an approach creates a true win-win situation for consumers, business, society, and the world. Plumpy'nut, the result of public-private partnership of Nutriset, a French private firm specializing in therapeutic food, and the Institute of Research for Development, a French public research institute based in Paris, has demonstrated the power of this collaborative involvement. The company makes money, and millions of lives are saved.

But even with government, business, and nonprofits working together, success will not become the norm until the *process* of innovation is defined and executed, creating an infrastructure for continuous successful social innovation. The mission of this book is to offer a framework that will engage all the sectors in a common quest to solve the world's most vital challenges.

The Social Impact Framework

The Social Impact Framework is put forth as a guideline to enable the social sector to develop innovative solutions using state-of-the-art innovation techniques that have been proven in the corporate world. The most vital aspect of this framework is the order in which the steps in the process is executed. People naturally want to solve a problem by cranking out ideas—lots of ideas—hoping to find that golden nugget that will work. Instead, research has shown that the best practice is to first separate the needs from the solutions, and then develop an understanding of the needs before a single idea is considered. Once we systematically seek to understand the needs, only then will we be well armed to generate solutions that will work.

The Social Impact Framework's six steps take an organization from investigation (identifying the areas of need), through ideation (devising the solutions and unique financial models), and finally implementation (delivering the solution). (See Figure I-2.) The good news is that we do not have to create this framework from scratch—all the elements already exist, have already been tested, and have proven to be successful.

As noted author Andrew Winston (*Green to Gold* and *Green Recovery*) indicated in an interview in 2009, "As much as we need . . . these giant changes in how we do things, the *process for getting there* . . . doesn't have

Figure I-2 Social Impact Framework

to seem alien"[7] (emphasis added). He in fact discusses the possibility of "systematizing" innovation by using techniques that achieved corporate success, such as that of Clayton Christensen, and applying them to the question of how the company can benefit society and specifically how it can make a "green" impact.

Implementing a process for systematizing innovation yields the greatest success. This book combines the techniques of several noted thought leaders in the innovation space including Clayton Christensen's Disruptive Innovation, Tony Ulwick's Outcome-driven Innovation, and Tim Brown's Human Centered Design. The objective of this book is to create a step-by-step process so that organizations can implement a systematic approach to innovation that will become a means of generating new solutions for decades to come.

Book Organization

This book is divided into three parts: investigation, innovation, and implementation. Part 1 Investigate, covers the first three steps of the process: defining the social challenge, identifying and prioritizing the unmet needs of the members of the ecosystem ("member needs"), and carefully examining the identified opportunities and their context. In these chapters, we explore the tactical steps and tools that can be used to understand a social challenge in detail.

Part 2 Innovate the Solution, covers the next steps in the process, offering practical tools and techniques for generating ideas, resolving contradictions, overcoming constraints, and developing a financial model to support the innovation.

Part 3 Implement the Solution, focuses on the development of business models and diffusion strategies for the innovation. This part also explores how the methodology can be used in several key areas of social concern today, including health care and natural resource management as well as the various ways this method can be used by different levels and organizations of government including education, military operations, disaster relief, and others.

Part 1: Investigate

Chapter 1: Define the Social Challenge

As stated in the literature on so-called wicked problems (that is, problems that are large, complicated, and intractable), one of the biggest challenges in solving wicked problems is defining them. The first three steps in our social innovation methodology define the problem in detail and with precision. First, we need to identify the members of the ecosystem. In the Plumpy'nut scenario, for example, the members include aid workers, health-care workers, parents of the malnourished children, and the malnourished children. We also need to clearly understand the context within which the social challenge exists. This includes the political, cultural, and social framework of the social challenge as well as the physical and human components that can cause solutions to fail. In step 1 we begin to sketch these out; we delve deeper into these issues in steps 2 and 3. The chapter concludes with the development of the Social Innovation Blueprint, a tool that will help frame and define the challenge to be addressed.

Chapter 2: Understand and Prioritize Needs

Next we need to understand the member needs including the jobs (tasks, goals, and objectives) the members of the ecosystem are trying to achieve, the steps for executing the job, and the constraints that prohibit members from doing the job effectively. Each of the ecosystem members has specific needs that are important and not well met by current solutions. In the Plumpy'nut scenario, for example, health-care workers needed to reduce the amount of space required in the hospital for feeding children. The parents needed to increase the number of days of nourishment that could be obtained at one time, ensure their children would eat the food, and so on. Aid workers needed to increase the amount of aid that could be delivered to a location in one trip and reduce the cost of transporting the food. All these needs must be identified in the early stages of the process, and they must be prioritized in order to create a breakthrough innovation.

In all social ecosystems, there are very real constraints to success—both human and environmental. In the Plumpy'nut scenario, environmental constraints to success included a lack of clean water and no refrigeration (Plumpy'nut is ready-to-eat and vacuum sealed), as well as political unrest, which made transportation to the affected areas difficult and dangerous (the small packages allow for much fewer trips to the clinic to obtain the food aid). Human constraints included different languages and low literacy rates (addressed by easy-to-open packages, and no mixing or preparation instructions needed). For the solution to be successful, all of these constraints had to be addressed.

Chapter 3: Examine the Opportunities

In this chapter, we explore the *context* in which the opportunities exist, including the solutions members currently use to get jobs done, why the opportunities have not been solved in the past, what kind of workarounds exist today, and so on. We also investigate the technologies that are being used to address the opportunity today as well as technologies that do a similar type of job in a different industry. For example, when looking at technology available to determine whether the patient's vitals are improving, we would explore solutions in the health-care industry that are already used for monitoring vitals at home; however, we will also look at other industries where remote monitoring is taking place such as the home security industry and the technologies they are using to address a similar job.

Finally, we want to understand the competitive environment to determine what solutions are in the market today. Because we are focused on the job, "competitors" include any product, service, or program that helps the member execute the job in the market today, including traditional competitors (such as companies in the same industry, direct competitors, and so on), as well as nontraditional competitors. For example, if we are trying to find competing solutions for determinining the cause of symptoms in the body, some traditional competitors might be the primary care physician, an urgent care center, or an emergency room. Nontraditional competitors might include do-it-yourself solutions such as WebMD or retail clinics such as MinuteClinic. It is vital to gather all of this information on the opportunities as this detailed, holistic understanding of the problem provides the necessary foundation for formulating innovations that will succeed.

Part 2: Innovate the Solution

Chapter 4: Devise a Workable Solution

Once the information from the first three steps is at hand, it's time to generate a solution. As mentioned earlier, to address social problems, a collaborative platform is best. In "Platforms for Collaboration" (*Stanford*

Social Innovation Review, Summer 2009), Satish Nambisan suggests that an electronic platform for this type of international collaboration is easily within reach with today's technology.

Once a platform is established, there are myriad techniques for generating creative solutions, resolving contradictions, and overcoming obstacles, only a handful of which we can cover in this book. There are many excellent creativity techniques that can be applied once we know where to target the creativity. When the needs are so carefully identified and their causes understood, the matter of generating ideas becomes much less complicated.

Chapter 5: Develop a Business Model

Successful social innovation requires addressing issues within the business model as an inherent part of the innovation. Traditional business models are concerned with two things—how will the solution make money and what will the costs be to deliver the solution. We expand the business model to describe how the concept will create, deliver, and capture *value* and how it will then give back to the world. When a customer uses the solution, there is the notion of "reaction." By using or consuming a solution, a reaction takes place in people, their environment, their community, and the earth. The business model must take into consideration the social value created and the environmental impact extracted.

For example, let's look at a car. We can use a car for transportation to and from work, and in doing so, we have things we *get* from this experience and things we *release* by this experience. In the *get* column, we get our desired jobs done—we get to work, obtain protection from the elements, have entertainment while waiting in traffic, and so on. In the *release* column, we release pollution and noise and contribute to congestion in the environment. Makers of cars could focus on working on the driving experience and making it exceptional—which they do. However, a shared value concept a carmaker would also be concerned about the impact the car has on the road, the community, and the planet. This is just now happening with the advent of quieter hybrid cars.

Let's look at the Plumpy'nut example. The product gives nutritious food to starving people at a low cost. The release of this product is the transportation to get the food to the people in need which was originally quite high when the production took place in a centralized location. True to the social value that this organization seems to espouse, they have moved to a model where they are making the food much closer to the people who need it most, thus reducing dramatically the effects of transportation.

This chapter also tackles the difficult question of how to get the concept funded and who will pay for the solution if the beneficiary of the solution cannot afford to pay. Currently, many innovations in the social space are launched and funded by nonprofits, government entities, and even for-profit companies. Although the Plumpy'nut product was developed for malnourished children, its salient characteristics (long shelf life, no preparation necessary, packaged in small units, and easy to carry) make it appealing for campers, for day-care providers, for disaster relief, for military rations, and so forth. In this chapter, we introduce several triggers that are helpful in identifying new business models.

Part 3: Implement the Solution

Chapter 6: Diffusion of Innovation

The diffusion of social innovations can be difficult and complex. Geographical, cultural, and social differences, as well as technical and environmental issues all come into play when the solution is extended from one location to another. In the case of Plumpy'nut, it was very difficult and expensive to deliver the product to certain inaccessible areas. The solution was to establish local "kitchens," where personnel were taught to make the Plumpy'nut supplement from local raw materials. Additionally, many solutions run into the "not invented here" syndrome and the new sites for the solution often feel that the innovation is being forced upon them without fully understanding their needs. In Chapter 6, we look at several issues that plague the diffusion of social innovations and how they can be overcome by applying an abbreviated version of the Social Impact Framework.

Chapter 7: Health Care

We look at the challenges facing the U.S. health care system and why the current platform and business model are creating a never-ending rise in cost and complexity. Through the application of the process described in this book, we explore ways that the scenario could be structured to yield interesting and beneficial insights to innovators.

Chapter 8: Conservation

In this chapter we look at how the needs of consumers and people around the world can be balanced against the needs of nature and the need to conserve our natural resources. The chapter explores the unique issues surrounding innovation in this critical space, specifically how to engage nature as a member of the ecosystem whose needs must be considered in order for the innovation to take place.

Chapter 9: What Citizens Want

In this chapter we focus our attention on the needs of citizens and provide a framework that government organizations can use to systematically identify the real needs of their citizens for purposes of innovation. Whether the issue is education, disaster mitigation, energy regulation, health and human services, or any other of the hundreds of jobs the various levels of government perform, officials and policy makers can benefit from understanding the unmet needs of the population they serve, leading to innovative solutions that will have the most impact across all members of the ecosystem.

A National and International Trend

More and more governments are forming innovation centers, suggesting a move toward a more methodical approach to social innovation. In the United States, President Obama recently formed the Social Innovation Fund, whose mission is to "catalyze partnerships between the government and nonprofits and businesses and philanthropists in order to make progress on the president's policy agenda."[8] The Social Innovation Fund will also be tasked with identifying innovative programs and solutions that are working well, which can be used as examples and be replicated. It is encouraging to see social innovation given such prominence in the president's agenda.

Elsewhere, Scotland has created a series of nationally sponsored technology institutes, which are areas of innovation focus. These innovation programs do more than merely provide capital for innovation: they conduct what they call market foresighting to identify where new market growth is likely to take place, and they then fund research and development to meet those projected future demands.[9] By focusing on the "precompetitive" point of innovation, the programs can get ahead of the market. One of the first efforts was to develop chronic wound care solutions, especially for wounds resulting from the complications of diabetes, a condition that affects hundreds of millions of people worldwide. Because the institutes relied on a solid methodology, they were able to create a point-of-care diagnostic platform that could be applied in either a clinic or a community setting.

Ireland also has a state-sponsored innovation program. Ireland's program focuses on helping indigenous industries identify new innovation opportunities and build a strong workforce with specialization in applied research techniques. It provides a way to transfer knowledge to the marketplace by helping companies with the commercialization process.[10]

While in many national innovation efforts the focus is on using science and technology to drive economic development, our hope is that more countries will follow the lead of the United States and Scotland and add a focus on social innovation—generating innovations that will serve the greater good of the people. If countries that are willing to support and foster innovation for purposes of wealth creation and economic prosperity were to add a parallel effort to identify innovative ways to provide health care, improve education, and end hunger, we might see great advances in these areas. Innovations that target the basic needs of a country's citizens are just as vital to the success of the country as innovations that promote a vigorous economy.

Tools and Templates

Throughout this book, we introduce several tools for executing the innovation process. These tools are available for download to those who have purchased either the hard copy or e-book by visiting our website at www.TheSocialInnovationImperative.com. I hope you find these tools valuable as you seek to apply the methods outlined in the book.

Part | 1

INVESTIGATE

Chapter | 1

DEFINE THE SOCIAL CHALLENGE

Poverty, hunger, terrorism, natural disasters, environmental damage, lonely elderly, poor graduation rates, inaccessible health care—these are issues we know well. They have plagued us for generations because they are "wicked problems."

Wicked problems are complex and involve several different constituents with competing objectives. They plague us because they defy our traditional means of problem solving: they are caused in numerous ways; they are interwoven and difficult to untangle. John Camillus, the author of the *Harvard Business Review* article, "Strategy as a Wicked Problem," observes, "Not only do conventional processes fail to tackle wicked problems, but they may exacerbate situations by generating undesirable consequences."[1] Wicked problems have no easily apparent answers; solving them can take generations. But there is hope. The key to solving wicked problems lies in *defining the issue with precision, clarity, and detail.*

These types of issues are also such that they cannot be handled by just one group, no matter how large and powerful. "Large-scale social change requires broad cross-sector coordination yet the social sector remains focused on the isolated intervention of individual organizations."[2] Successful programs are often found where coordination among the government entity, nonprofits, and corporations takes place.

Understanding the Ecosystem

Within any given social scenario—education, health care, resource conservation, or hunger alleviation—there are many different groups of people involved. These groups are highly interdependent on each other, each having its own set of needs. Thus, to define a wicked problem, the first step is to map the members involved and what they do. The groups of people working toward the overarching goal of the social scenario (for example teachers, students, and parents within the education scenario) comprise an *ecosystem*, and the groups of people within the ecosystem are referred to as *members*.

In social innovation, the members of the ecosystem are the customer for whom we are trying to create value and improve satisfaction. The challenge is that while there are some needs members all agree on, there are several needs that are conflicting so that creating value for one group may detrimentally affect the needs of another group. These conflicting needs are often the source of the problem within the social scenario and a key part of the instability and dissatisfaction of the ecosystem.

Let's look at an example of an ecosystem in the education space. All the members of the ecosystem share the same overarching goal, "create educated, self-sufficient citizens"; however, some of the needs may bring them in conflict with one another. For example, students may be trying to accomplish the need of learning in a way that feels comfortable to them, but this need may be in conflict with teachers' need to maintain an orderly classroom and provide standardized content to large numbers of students. A true innovation helps members of an ecosystem resolve these conflicts and enables all members to meet their needs without too much impact on other members.

Further evidence of the conflicting needs among members is found in our health care system. The needs of the patient, the health care provider, and the payer are in serious conflict. Patients and payers are putting the squeeze on physicians to reduce their fees. Physicians who are faced with extremely large student loan debt coming out of school find that they cannot make enough money to justify the long hours, the school debt, and so on. This disharmony has driven physicians to simply give up their practices. In fact, recent surveys show that over 10 percent of physicians plan to quit their practice, when there is already a shortage of some types of physicians. Such is a typical result with a severe case of conflicting needs within the ecosystem. *The goal of social innovation is to maximize the satisfaction of all members of the ecosystem with new solutions that will address the needs across the spectrum.*

Jobs: A Simple Shift in Perspective

A great deal of success within corporate innovation has been a result of gaining clarity concerning what is generally referred to as the "fuzzy front end." The front end of innovation involves understanding the problem, identifying the customer needs, as well as the constraints that must be overcome. Elimination of this fuzziness has been achieved as a result of a simple but elegant paradigm shift in the way organizations view customer needs and the timing of obtaining those needs. The introduction of using "jobs-to-be-done" as a standardized method of defining needs and the adoption of a "needs-first" approach have made substantial improvements in the innovation process.

Jobs are defined as the goals and objectives that people want to accomplish or what they are trying to prevent or avoid. In the commercial innovation literature, jobs are what motivate people to buy a product or service such as an iPhone, which enables them to "be productive while on the go," or auto insurance so they can "protect against financial loss in the case of an accident."

In the social space, jobs also reflect what people are trying to get done and what motivate people to engage in the activities they do. For example, students want to prepare for a future career, aid workers for the Red Cross want to provide supplies to those who are displaced in a natural disaster, and physicians want to educate patients on how to improve their cardiac health. Jobs explain why people help or do not help others, what goals they want to achieve, and what they are willing to do without.

In the sphere of social innovation, breaking down social problems into the jobs that the members are trying to get done allows us to identify where solutions are needed and what constraints are preventing the successful execution of that job. Table 1-1 shows several *sample* jobs of different social scenarios.

Table 1-1 Sample Jobs from Various Social Scenarios

Social Issue	Who	Sample Jobs
Resource conservation	Rural farmers	Maintain a consistent crop
		Improve the production of crops
		Avoid depleting the land
		Provide jobs to locals
		Etc.
City budgets	Council member	Determine the priority of the citizens' needs
		Determine how to allocate limited dollars
		Determine the economic impact of cutting an area of the budget
		Etc.
Hunger	Parents	Determine where food is available for the money they have
		Determine where the children can get a meal
		Obtain transportation to the location where aid is available
		Acquire food that is available
		Determine how long that food can last
		Distribute food among the family members
		Etc.

Childhood obesity	Physician	Identify whether the child has a weight problem
		Identify the child's eating habits, e.g., where the child's calories are coming from
		Educate the child on better eating habits
		Educate the parents on the causes of obesity
		Educate the parents on the impact of obesity on the child's health and well-being
		Etc.

Consider the results in the corporate world, where analyzing jobs-to-be-done has led to some breakthrough solutions—like the iPhone. The iPhone's focus is all about helping customers achieve the jobs they want to get done while they're on the go. While the primary job of the phone is to communicate with others, there are a lot of *other* jobs that people on the go want to get done as well. People want to find a restaurant or a Starbucks near their current location. They want to find out what movies are playing near them. They want to communicate with several people at a time. They want to be entertained in short periods of downtime. In fact, at last count, more than 200,000 applications ("apps") have been created for this amazing device to address specific jobs that people want to get done. An *app* is a very job-specific program that, instead of executing myriad jobs, simply executes one specific job very well and for a very low cost.

The global success of the iPhone and apps package testifies to the universality of the concept of the job. The beauty of making the job the unit of focus is that it brings the discussion to the most basic level, that is, "be productive on the go" and steers away from preexisting or preconceived notions of solutions, such as simply making a better cellular calling experience. Such definition and clarity can make social innovation significantly more effective.

The second major improvement to the success rate of innovation activities is the focus on a needs-first approach. The needs-first approach is one that identifies the needs of the customers, or ecosystem members, and prioritizes them before any ideas are generated. With this approach, the idea-generation activities are highly targeted on the most important and unsatisfied needs. Contrast this with what has been the traditional ideas-first approach where ideas are generated by people within the organization. They are then put through a process of gaining customer reaction, adjustments are made, again customer feedback is obtained, and so on. This method involves a lot of trial and error and is highly inefficient, yet it was the norm for a long time. In fact, many product development methodologies actually start with an idea and don't get to the customer needs until well into the process. By simply moving the needs gathering to the front of the process, significant improvement can be made in the success rate of innovation.

Even Thomas Edison, the world's most prolific innovator who spawned the creation of six industries, adapted a "needs-first" approach early in his career after an initial failure with an "ideas-first" approach. He began to define success in terms of "utility" which he defined as "the ability to satisfy a customer need or marketplace need."* He realized that it is easy for people with talent for creating new products and services to get caught up in clever solutions without thinking about whether they had a real need in the market. Given the speed at which we need to generate solutions in the social space, it is vital to stick with a plan that works and has been shown to be effective—a "needs-first" approach.

Disruptive Innovation as a Strategy for Social Innovation

Clayton Christensen's disruptive innovation has significant application to social challenges. "Disruption is a positive force. It is the process by which an innovation transforms a market whose services or products are complicated and expensive into one where simplicity, convenience, accessibility and affordability characterize the industry."[3]

In the social space this concept is critical for bringing solutions to those who often need them the most, to groups of people that have little or no access to the solutions that are available—new technology, services,

---

*Sarah Caldicott-Miller, "Ideas First or Needs First: What Would Edison Say?," Whitepaper, 2010.

products, or programs—even though they have unmet needs that could be addressed with these solutions. In the *Harvard Business Review* article, "Disruptive Innovation for Social Change,"[4] Clayton Christensen and his colleagues argue that social innovation should be targeted at the needs that are *overserved* by existing solutions and those individuals who do not have the wealth, access, or skills to acquire the solutions that exist today— the nonconsumers (or nonusers).

This group is a significant size in most cases because organizations tend to focus on addressing the needs of mainstream customers who want more complex and expensive solutions; these solutions are often out of reach for a large sector of the population. This group of people, *nonusers*, are defined as people who "face a barrier that constrains their ability to solve an important problem. They must either go without or attempt to solve the problem to the best of their ability using existing products or services."[5]

Christensen identifies four distinct types of constraints that cause non-use, and we can find examples of all of them in social problems:

Wealth-Related Constraints

The most obvious constraint in many social scenarios is that the users simply cannot afford the solutions. For example, there are people who try to manage a chronic health problem but who cannot afford the medication that will help keep their condition in check. There are also parents who know that their child needs additional mentoring in a given school subject but the parents lack the money to pay for a tutor. Many victims of natural disasters don't have insurance or the means to rebuild their home and livelihood. This type of nonconsumption is a very real issue—how do we provide access to good solutions for those who cannot afford them?

More and more organizations are beginning to make products and services available to the vast population at what is known as the *bottom of the pyramid* (BoP). C. K. Prahalad, professor of Corporate Strategy at the Stephen M. Ross School of Business in the University of Michigan, now deceased, has demonstrated that ". . . by virtue of their numbers, the poor represent a significant latent purchasing power that must be unlocked."[6] The good news is that much attention is being given to creating solutions for this group, resulting in profits for the companies pursuing this market and helping the world's poorest meet their needs.

It is important, however, to go through the process of understanding which jobs are important for the people at the bottom of the pyramid. Many a company has been burned by making a product available to that group, just to find out that its members were nonusers because they had no interest in getting the job done that the product or service provided. "Companies with brilliant track records of marketing to the developed world have failed to launch high-impact projects, such as Nike's disappointing introduction of the $15 World Shoe. Even more striking is the example of Procter & Gamble—one of the world's best marketing and management firms—which, despite the identification of a 'market need' of clean water, has been unable to successfully market their PUR water purification technology to the BoP. Indeed, not a single corporation has created a viable clean water business at the BoP, though there have been many attempts."[7] The suggested reason for these failures is that because there is no product market, there is no way to judge demand. We would argue the opposite. In markets where there are no products, that is exactly where the jobs-to-be-done methodology excels. In these markets it is *imperative* to understand the jobs that these members want to get done and the prioritization of these jobs. Only then will organizations know which products to pursue for this group of people. These data alone define and quantify the demand of that market which even in the absence of existing product because we are measuring demand of the job. If we really want to sell to the BoP, we must understand jobs they want to get done. Too often assumptions are made that are wrong.

Skill-Related Constraints

These constraints prohibit a group of potential users from using a product or service because they lack a necessary skill to do so. An example is found with seniors who are unfamiliar with computers, and this often prevents them from being able to access information and services available on the Internet. We also see this with children who have learning disabilities as they struggle to get the same education as their peers.

Skill-related constraints require innovations that make the job easier or that allow the job to be done on a simpler platform that can be used by less skilled people. An example is the introduction of new medical technologies that allow patients to do more medical activities themselves, from monitoring blood sugar to taking blood pressure to even checking for ear infections. Another good example is a program in Texas called The Green Corn Project in which volunteers with skill in agriculture and gardening help the poor to establish a small community garden to provide healthy food for themselves and their families.

Access-Related Constraints

Access-related constraints have to do with location or context. For example, people who do not have access to television are more likely to be unaware of world events. However, the "Internet has played a powerful role in democratizing access to information"[8] allowing more and more people to participate in events that concern them. Through services such as Wikipedia, blogs, YouTube, and others, mainstream people can now publish their thoughts, bringing voice to countless more people than ever before. We've seen the amazing impact this has had during the Arab Spring of 2011 where these tools helped to overthrow repressive governments throughout the Middle East. The distribution of information about the ruling party and how people could become involved in an uprising resulted in the end of a 30-year authoritarian rule in Egypt.[9]

Another great example of social innovation that overcomes the access constraint is a newly developed center for victims of domestic violence. These new centers were created in response to data that showed that "most of the women that die in domestic violence in America die after they've sought a restraining order, after they've called the police, after some interaction with the system."[10] The first center, established in San Diego in

2002, brings together police, prosecutors, social service agencies, and non-profit advocates to assess the immediate danger of the victim, assist in the next steps, and provide the support needed to ensure the victim's safety. Since its inception, the city has seen a "90% drop in intimate partner homicides."[11] The cross-agency collaboration, the wider range of services, and the improved public-private cooperation has created truly amazing results simply by improving access to services that already existed.

Time Constraints

The last type of constraint that causes nonuse is that of time. This occurs when a product or service is too time consuming and cumbersome to use. Former users are a good source of finding out whether there is a time constraint that is prohibiting adoption of a given solution. Later in this chapter we look at MinuteClinic which is an example of disruptive innovation which has overcome time and access constraints.

The disruptive innovation strategy is highly effective in addressing social issues. By exploring needs of the nonconsumers involved in the social scenario, the focus can shift to how to create a solution that is simple, affordable and easily accessible to the nonuser. Nonusers lacking either skill or wealth need a solution without the complexity and all the extra features found in the primary product or service. Solutions that get the primary job done "well enough" are all that is needed. The more that the products, services, or programs can be made to be self-administered or self-directed and the less complex, the better the chance of overcoming issues of nonuse. All in all, the disruptive innovation strategy should be one of the first perspectives to be considered when attempting a social innovation.

Defining the Social Scenario

The scenarios we target for social innovation are those in which a change will benefit more than just a single group of people. The goal of social innovation is to improve our society, the way we live, the way we interact, the way we care for people, and so forth. Before we can innovate, however, we must first define the social scenario—the scope or framework within which we will innovate. We must define the goal of the innovation initiative, learn who is involved, identify the scope of nonconsumption and the reason for it, and understand the status quo, especially the current "rules" and the biggest issues.

Let's take the case of health care delivery. The health care system includes a broad range of jobs that can easily be addressed. There is a large group of nonconsumers of traditional health care products, such as insurance coverage, many of whom use government programs as their solution. So as we look at this situation, we need to break it down into a reasonable social innovation initiative. There are several ways to narrow the scope:

1. Zero in on a specific demographic (e.g., uninsured people, obese children, people in a specific geographic region).
2. Select a type of job. For instance, in health care, one could choose to focus on chronic illness, innovations in payment methods, or service delivery in an in-patient setting.
3. Focus on a special-case instance in the scenario. For example, one might choose to innovate around health care during a community crisis such as a natural disaster, a large accident, or an epidemic outbreak.

Scope management is important to ensure that the innovation efforts will have a meaningful impact and that the solutions will improve the situation. To test whether you have defined the scope appropriately, determine whether you and your team will be able to create solutions within the scope you have defined. Will the results of your efforts provide you with direct insight into what you can create (e.g., products, services, programs, etc.)? Later in this chapter, we go step-by-step through a framing tool designed to help direct you and frame your initiative.

Below are several scenarios that my colleagues and I have been actively working on in the social sector:[12]

1. *A program for the K–12 age group in education, focused on how children learn new material.* The members being studied are students, parents, teachers, and other content managers within the school system (e.g., librarians).

2. *A large initiative to identify sustainable commercial building practices.* The goal of the corporation that is undertaking this initiative is to develop innovations within their existing product lines for managing natural resources, as well as process innovations that result in a more sustainable business practice. The members being studied include commercial building architects, builders and contractors, facility managers, and occupants.

3. *Health care delivery initiatives in hospital service areas around the country.* The targeted jobs include treating patients and identifying services that are needed, and the goal is to find innovative ways of delivering care. The members being studied include physicians, patients, employers, and hospital administrators.

4. *A program to help elders transition from independent living to a nursing home or living with a loved one.* Members include the older adults, their caregiver/family, physicians and nursing home personnel, and government agencies that work with the elderly.

As we look at a social scenario, we need to consider the four key pieces to the puzzle: the *members* of the ecosystem, the *platform* through which jobs currently get done, the *needs (jobs)* of the ecosystem members, and the *constraints* of the ecosystem. Members and platforms are discussed in the following pages; needs and constraints are explored in Chapter 2.

Members of the Ecosystem

Throughout the methodology introduced in this book, you'll note that we refer to the scenario in terms of the challenge and the *ecosystem*. In nature an ecosystem is a group of entities that work together and rely on each other for their survival. In social innovation scenarios, this description also accurately portrays the interactions, connections, and interdependencies that exist among the groups involved in the social scenario. Consider social issues such as education, health care, caring for the elderly, feeding the hungry, creating job growth, conserving natural resources, and so on. All these issues involve numerous entities from all sectors—government, nonprofits, Non-Government Organizations (NGOs), and corporations. Throughout the book *members* is used to refer to the groups that actively participate in the ecosystem in order to generate value or receive value.

Most social scenarios involve at least three types of members, each with its own distinct set of needs. These three types of members are the job executor (primary or secondary), the job beneficiary (primary or secondary), and the job overseer (or third party). We discuss the role of each of these further.

Job Executor

In all scenarios, there is at least one primary job executor. Those in the primary job executor role work most closely with the receivers or beneficiaries of the program or service and have direct responsibility for ensuring that the beneficiary's job is accomplished. Take education, for example: The teacher is the primary job executor, and the student is the beneficiary.

In most social scenarios, there are more than one executor, who support the primary job executor in helping beneficiaries get their job done. For example, in health care, the physician is often the primary job executor, with the nurse, physician's assistant, x-ray technician, and many others serving as secondary job executors. In a disaster relief scenario, the primary job executor is often a first responder or group of first responders with secondary executors including the fire department, police officers, the Red Cross, and community- or church-based support organizations. Secondary job executors may be critical members of the ecosystem. In the health care ecosystem, for example, nurses and physician's assistants play a major role in preparing the patient for treatment. In education, parents play a significant role in helping the child get school work done. In government, there are many secondary job executors who help elected officials serve the citizens. These secondary job executors not only have a role in the focal job, but they often help with several related jobs as well.

Job Beneficiary

Although the description of the job executor's role may make the job beneficiary sound like a passive recipient, in fact job beneficiaries are often active members of the ecosystem. The job beneficiary is the member of the ecosystem who the ecosystem is set up to help and support: in an education ecosystem, it is the students; in a health care ecosystem, it is the patients. As with job executors, there can be more than one job beneficiary. Other beneficiaries often have roles that support the primary beneficiary.

Interestingly, in many social scenarios, a great deal of the disharmony occurs when the beneficiary is not fully engaged in his or her job. There are dozens of examples of apathetic job beneficiaries: citizens who don't vote, students who won't study, victims of disasters who do not heed the evacuation warning. Their apathy creates major problems within the social scenario. All members of the ecosystem must actively work to execute their jobs in order to achieve a stable and functional ecosystem.

Let's look more closely at what happens when beneficiaries do not try to get their jobs done. In the health care scenario, the patients' main job when they are not sick or injured is to maintain the body's ability to function (e.g., provide nutrition, get exercise, avoid harmful substances). Yet a 2010 Webmd.com survey shows that 61 percent of Americans are obese or overweight, over 20 percent of men and 18 percent of women still smoke, and only 51 percent claim to exercise three or more times a week. In other words, the beneficiaries are *not* getting *their* jobs done—by choice.

Physicians are extremely frustrated with patients who refuse to take care of themselves and then simply turn to the health care system to fix them when their bodies fall apart following negligence or misuse. In 2004, we found that patients' attention to their own health maintenance was among physicians' top unmet needs—a measure that showed that they were having trouble meeting their job of caring for patients' health. This trend of the beneficiaries' failure to get their job done has caught the attention of other members of the ecosystem, and action is being taken to ensure that the job beneficiaries are held somewhat responsible for performing their job. Customers of health insurance are being charged higher rates when they choose to engage in lifestyle issues that can affect their health such as smoking and in some cases obesity. The state of Arizona has implemented the "stupid driver" law. Drivers who ignored barricades and driven into flooded roads resulting in a rescue operation are now being held responsible for paying for the rescue. Unfortunately, when job beneficiaries abdicate their responsibility for doing their jobs, the consequences—financial, or worse—will catch up with them.

Overseer or Third Party

The overseer or third party is another member of the ecosystem that includes a number of different types of entities such as regulators, supervisors, suppliers, partners, administrators, payers of services, policy makers, and so forth. The third-party members in a health care ecosystem would include hospital administrators, insurance companies, and government regulators. In an education ecosystem, they would include district superintendents, government regulators, and teachers' unions. By definition, these types of members are likely to cause discontent in the ecosystem because their jobs include exerting controls, influence, and restrictions on the execution of the overall job. The regulations and constraints that the job overseers are charged with enacting may impede other ecosystem members' ability to get their jobs done.

Although the rules that the third parties impose are often to protect the members of the ecosystem, those rules are often seen as undesirable by the job executors and beneficiaries. In the health care ecosystem, these third parties may be perceived as withholding care or hampering the physician's ability to practice medicine—even though they are also making sure that the consumer is being cared for, preventing fraud and abuse of the system, and making sure that the system remains solvent. The general public often doesn't see these goals as valuable, especially when these goals are contrasted with stories of denial of service or burgeoning bureaucracy. However, third parties are important parts of social innovation scenario ecosystems and cannot be ignored.

These members are critical to study as they provide key insights as to the dynamics and connections within the ecosystem—the "rules," an understanding of why things are the way they are. Most importantly they provide insight as to the criteria that they use to determine the best solution to address the problem.

In some cases, social innovations improve the ecosystem in ways that remove the necessity for third parties or overseers to be part of the ecosystem at all. For example, in Texas, Scott & White is one of the nation's largest health care systems with 10 hospitals and over 1,100 physicians throughout central Texas; however, it also fills the role of the payer through its Scott & White Health Plan, selling directly to employers and individuals and offering standardized health plans, thereby eliminating the insurance company and its requirements from the ecosystem.

Table 1-2 defines the roles of the three types of ecosystem members in several social scenarios.

Table 1-2 Roles of the Members of a
Social Innovation Scenario Ecosystem

	Job Executor	Beneficiary	Overseer or Third Party
Health care	Physician: treat the patient	Patient: execute treatment plan	Insurance company: pay for treatment
Education	Teacher: provide new content to students	Student: learn new content	Parent: support the student in learning new content
Government	Elected officials: manage services for the city, state, or federal government	Citizens: utilize services from the city, state, or federal government	Vendors: provide services to the citizens of the city or state
Disaster relief	Relief workers: treat victims	Victims: follow the instructions of the first responders	Aid agencies: coordinate efforts of relief workers

Sustainable building	Building operations manager: manage water and energy usage	Tenants: utilize water and energy	Architect, builder, etc.: design water and energy systems for the building

The purpose of mapping these roles within the ecosystem is to identify the members who are the most critical for purposes of innovation. This is especially important in managing the scope of the initiative. In practice however, do not be too concerned about whether you have labeled a member a job executor or an overseer, as long as that member's core job, and the relationship between all the members, is clear.

Other Stakeholders

The final set of inputs we need to consider are the criteria of the key stakeholders—the groups sponsoring the innovation plan. These stakeholders are likely to be executives within your organization or may comprise individuals from other organizations, corporations, nonprofits, and even government agencies who have committed themselves to the innovation program being undertaken. Thus we must understand their criteria for what is considered a successful idea. For example, if the key stakeholders want speed to implementation more than they want low cost, we need to know that. The stakeholder criteria are vital in directing the ideation process. Thus while capturing the needs of the ecosystem members, the process of obtaining the stakeholder criteria can begin as well. Table 1-3 shows a sample of a stakeholder criteria rating sheet. The goal is to understand the prioritization among the stakeholders of these criteria.

In practice we find that the most effective approach (and the one executives will actually complete) is a simplified high-level category rating. For this version, the rater is given 100 points to allocate across all the categories, indicating their priorities of the concepts to be generated. The average point assignment can then be calculated and to provide the prioritization of criteria. Obviously, stakeholder criteria requesting maximized societal

Table 1-3 Stakeholder Assessment Tool

Categories	Organization 1		Organization 2		Organization . . . n	Prioritization
	Rater 1	Rater 2	Rater 3	Rater 4	Rater . . . n	Average
External cost to customer must be low						0
Effort—internal time to develop must be low						0
Internal cost to develop must be low						0
Risk—concept must have low technical, political, and organizational risks						0
Sustainable advantage—ability to own and protect what is created						0
Societal impact—value created for society is high						0
Total	0	0	0	0	0	

value and speed to launch is going to yield much different concepts from those that are focused on sustainable advantage and low customer cost. These criteria are referred to again during the ideation discussion.

Platforms

After understanding the roles of the members of the ecosystem, we turn our attention to the platform on which those jobs are executed including: the infrastructure, subsystems, processes, technology, and people who are involved in executing the job or executing the social scenario. For example, the iPhone—and more recently the iPad—is the platform upon which Apple's customers accomplish their multitude of jobs related to finding information, communicating while on the go, and being entertained. It is important to understand what platform or platforms are currently involved in one's targeted social scenario because in many cases the platform is part of the problem. It may be insufficient, inefficient, or incapable of handling ecosystem members' new jobs. Too often, when organizations try to solve wicked problems, they fail to take the platform into account: it is seen as a given or constant, when in fact it, too, is an area of potential innovation, often the greatest opportunity for break through innovation.

The situation with the health care system is a great example. Legislators are attempting to solve the health care crisis by tweaking portions of the existing platform, when in fact the platform cannot handle the jobs that patients need to get done. Instead, the legislators should first identify the key jobs that the members of the ecosystem are trying to get done and then begin to innovate—platform and all.

The MinuteClinic demonstrates the value of introducing a new platform to address specific jobs such as obtaining treatment for a common ailment, such as a cold, flu, ear infection, and so on. Instead of trying to address this job on the traditional platform (treatment by primary-care doctor or at the emergency room), MinuteClinic has created a platform that is much better-suited to address it: specialized clinics, staffed by nurse practitioners who only treat patients with common ailments, eliminating the physician from the model and yet the job of treating the patient is still accomplished. Let's look at the advantages of this new platform. First, it takes a lot of pressure off primary-care physicians by siphoning off those patients who likely need only an antibiotic or other minor medication. It also provides an alternative to the emergency room for patients who have these issues come up when their primary-care doctor's office is closed (nights, weekends, holidays, etc.). It is much more convenient for patients who will not have to wait at their doctor's office to be squeezed in to an already very busy schedule or at an urgent-care clinic or emergency room. Finally, the cost of a visit to a MinuteClinic is often not much more than the typical insurance copayment at a doctor's office.

How can MinuteClinic afford to do this? Because the platform is designed precisely for the intended job—and nothing else. MinuteClinic's facilities do not have to purchase expensive equipment on the off chance that someone will come in with an injury that requires an X-ray—it doesn't address that job. It doesn't have to have specialists on call because it doesn't address that job, either. Table 1-4 illustrates the differences in the two different ways of addressing health care jobs.

Table 1-4 How MinuteClinic's New Platform Helps Physicians and Patients Get Their Jobs Done

Member	Need	MinuteClinic Feature	Result
Job executor (physician)	Reduce time spent with patients whose needs could be met by less skilled providers	Staffed by a nurse practitioner or physician's assistant	Fewer patients with minor issues fill up the physician's schedule thereby allowing time for the physician to focus on more complex problems
Job beneficiary (patient)	Reduce the time it takes to be seen by a provider from the onset of symptoms	No appointment needed—walk-in	Patient receives quick treatment for minor ailments

This is a prime example of how to carry out social innovation. To make a significant dent in the health care system's problems, we have got to start looking at the myriad jobs people are trying to get done and grouping the jobs together in such a way that platforms can be created or optimized to address those jobs. We explore the health care system in more detail in a later chapter.

Another area that requires focus on platform innovation is when nonconsumers are being targeted. In such cases, the platform must be simplified, less expensive, and easier to access. Take the Tata Motor's new "people's car" which sold for only a couple thousand euros. The platform of the car had to substantially change in order to accommodate this critical price point. While it still has the basic platform of four seats and four doors, it contains only a 600-cubic centimeter rear-mounted diesel engine with 30 horsepower, and it will use more plastic parts that can be glued instead of welded.

Needs

There are several different types of needs that we must explore in order to gain a full understanding of the members' needs. Figure 1-1 illustrates the types of needs and how they interact. We begin with the overall goal or objective that the member is trying to accomplish—the "job"—which we express in the form of a job statement. The structure of a job statement, developed by Anthony Ulwick in *What Customers Want* is shown in Figure 1-2.[13]

Focal Job

The *focal job* is the job of primary interest for a given member of the ecosystem. For instance, if we are interested in solving problems in the health care delivery arena, then the focal job of the *patient* is likely to be "obtain health care services." Jobs are always specific to particular players within the scenario—that is, each member of the ecosystem will have a different focal job. Thus in the health care example, the *physician's* focal job is to treat the patient's condition, the *hospital administrators'* focal job is to manage the day-to-day operations of the hospital,

Figure 1-1 Types of Needs

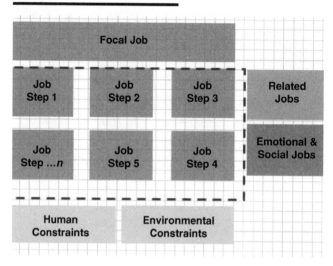

and the focal job of *benefit managers of employers* is to manage the health care benefits provided to the employees. Thus the focal job is specific to (a) the social scenario and (b) the member of the ecosystem.

Job Steps

Once we have the focal job identified, we then delve into the job to identify needs that the members have as they try to get the focal job done. For instance, in the course of treating the patient, what steps within that job are physicians struggling with? Where do they need new solutions within the focal job? In order to get this type of detail, it is necessary to "map"

Figure 1-2 Diagram of a Job Statement

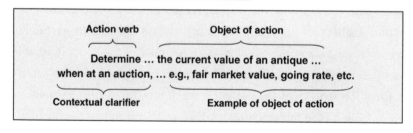

Figure 1-3   Universal Job Map

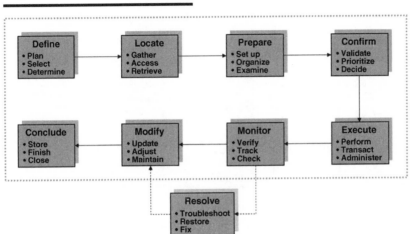

the job to understand the needs within it. The job map provides a detailed view of the execution of the focal job and gives insight into the areas of the job that are problematic.

In studying hundreds of jobs over the last decade, Dr. Lance Bettencourt, author of *Service Innovation* and a former colleague of mine, began to notice specific patterns that seemed to be present in all jobs, whether the job was to perform joint replacement surgery, manage personal finances, or manufacture paint products. Through his subsequent investigation of these jobs, he identified a set of common steps and formulated a universal job map that can be used as a framework for exploring any job (see Figure 1-3). This framework gives us a starting point from which to better understand the job, to guide the discussions with the members, and to ensure that we have truly captured the entire set of needs in the job. Additionally, it is in the creation of this job map that we capture the next set of needs—the job steps.

There are eight job step *categories* in the universal job map: *define, locate, prepare, confirm, execute, monitor, modify,* and *conclude* as well as suggestions under each of these categories for additional probing.[14] For example, within the define category, the steps will contain what must be done or understood before the job can be executed. These include such actions as

Figure 1-4 Job Map for Receiving Health Care Services

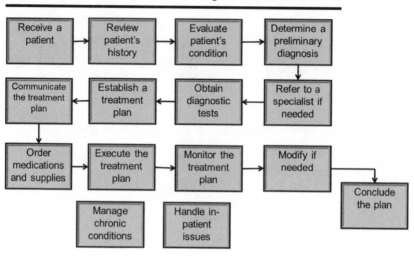

determining the objectives, planning how to execute the job, identifying what will be needed to execute the job, and so on.

The job map in Figure 1-4 outlines the steps within the job map for physicians treating patients. During the course of interviews, discussed in Chapter 2, the detailed steps are captured, rounding out the picture of the needs within the focal job. Within the step of "evaluate patient's condition," for example, there are several detailed job steps including:

1. Establish rapport with the patient, e.g., reestablish the connection, establish trust, etc.

2. Understand the patient's complaint, e.g., symptoms, reason for visit, etc.

3. Determine how the patient's condition has changed since the last visit.

4. Question the patient about the condition, e.g., when it started, what else has been affected, etc.

5. Avoid missing critical information, e.g., patient withholds information, too rushed to ask all the questions needed, etc.

6. Identify nonverbal cues that may help evaluate the patient's condition, e.g., anxiety, stress, nervousness, etc.

These detailed job steps are the next level of needs within the focal job. They are the needs that we use later to understand where in the process the member is struggling to get the focal job done.

There are several practice tips to consider regarding this set of needs. First, these job maps are very distinct from a process flow where the goal is to identify what is happening in a given process. Instead, the job map is designed to identify at each step what the member *is trying to get done*— not what the member is doing. This distinction may seem trivial but in actuality it makes the difference between a value map that can identify unmet needs and a process map that simply documents actions.

Let's look at a few examples. If, in the course of performing surgery, an anesthesiologist checks the monitor, we would not document "check the monitor" as the job step, even though that is the action that took place. What we are interested in is *why* the monitor is being checked. It turns out that the monitor is being checked to "determine whether the patient's vital signs are within an acceptable range". This, then, is the job step we would document. Consider the difference between the two statements. "Check the monitor" assumes that a monitor must be present, and it assumes that the anesthesiologist needs to or wants to check the monitor, when in reality he or she simply wants to know if the patient's vital signs are in range. True innovation will be found by helping the anesthesiologist "determine whether the patient's vital signs are within normal range," which may or may not involve making it easier to check the monitor. Focusing on current activity, rather than on the goals of the activity, is what limits innovation and is what we avoid by focusing on the goals.

Second, to ensure that the needs are stated in a way that will be free from solutions, stable over time, and so on, they should adhere to the standard framework of a proper job statement as defined earlier in Figure 1-2. Similar to jobs, the job steps within the categories are always free of solutions. By avoiding solutions in the job map, we can be certain that the job is being understood at its most basic level, ensuring that we are capturing needs on the job and not on any given technology or solution.

Third, in working on an innovation program, it is possible to begin to map out the high-level job steps. In fact, in many cases, you or members of your team may be or have been a member of the ecosystem (i.e., we've all been patients); as such, it may be possible to begin identifying steps you believe are part of the job map or related jobs list. When ready to delve deeper into the job steps, however, always conduct interviews with the actual ecosystem members to obtain confirmation of these needs. We talk more about how to gather these need statements a little later in this chapter.

Let's return to the Plumpy'nut example discussed in this book's Introduction and break down the focal job for the parent of a malnourished child. For the first universal job categories, there are numerous job steps listed. The remaining categories, in a real situation, would have as much detail as these first three do. (See Table 1-5.)

Table 1-5 Job Map for a Parent

Focal Job: Provide Nutrition to Children

Job Category	Job Steps
Define	1. Identify what foods are necessary for the child to maintain the desired weight
	2. Determine where this food can be obtained
Locate	3. Obtain transportation to the location of the food
	4. Determine how much food is needed
	5. Determine how much is available per child
	6. Obtain the foods that are needed for the child
	7. Make the return journey home with the food
	8. Avoid problems on the way home with the food, e.g., food spoils, is dropped, is stolen, etc.
Prepare	9. Determine if the child is well enough to eat
	10. Determine how much of the food to give the child at this feeding
	11. Determine how to make the food last for the necessary period of time
	12. Prepare the food for the child to eat, e.g., remove from packaging, make ready to eat, etc.

Confirm	13. Ensure the food is acceptable to eat (e.g., prepared correctly, not too hot or cold, etc.)
Execute	14. Give the food to the child
Monitor	15. Determine if the child is consuming the food
Modify	16. Make adjustments to the child's diet to ensure that the child continues to eat the right food
Conclude	17. Ensure that the child maintains the desired weight; remember what helped the child achieve the desired weight

Each of the job steps listed will become the core set of needs under the focal job, and there are many more under each of the job step categories.

Desired Outcomes

Desired outcomes, as defined by Anthony Ulwick, are the detailed metrics that help us understand specific issues within each of the job steps. The components of an outcome statement, as defined in Chapter 2 of *What Customers Want*, are structured as shown in Figure 1-5. The format in the figure ensures that the statement is written in a form that is (a) unambiguous, (b) free from existing solutions or technology, (c) timeless, and (d) specific enough to know whether an innovative idea has addressed the need or not.

Outcomes explore three specific areas of quality—speed, stability, and output. In general, as the members of the ecosystem attempt to execute their job, they are trying to get that job done well, fast, and reliably. Therefore, as we explore each step in the job map, the desired outcomes, along these three dimensions, are captured.[15] The detail available from

Figure 1-5 Diagram of an Outcome Statement

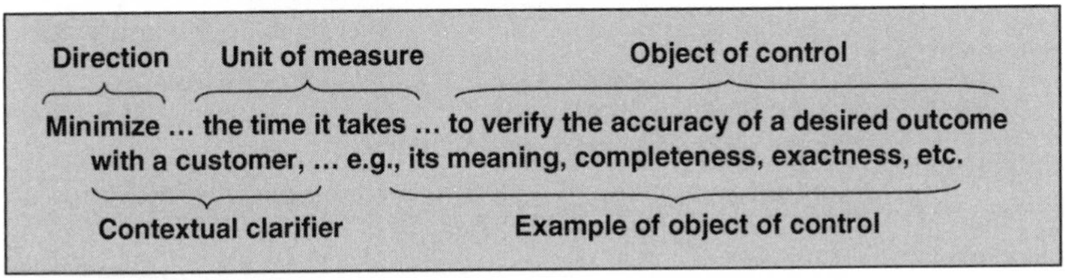

the outcome statements serve as common language for understanding the issues in depth.

Table 1-6 gives examples of outcome statements for the "execute" stage of the complementary jobs of delivering and obtaining health care and takes the needs to the "desired outcome" level of detail. Both of these jobs are executed during the health care delivery *experience*. When we are assessing an experience, we are breaking down the focal job of each member involved in the experience and investigating the corresponding job steps or outcomes.

Table 1-6 Outcomes within the Experience of Delivering Health Care during the Execute Stage of the Members' Job Maps

Job Map Category: Execute	Physicians: Job Step: Establish a Treatment Plan	Patients: Job Step: Implement the Treatment Plan
	1. Increase the likelihood of selecting a treatment plan that the patient will stick with (e.g., take necessary medications, do required activities, etc.)	1. Minimize the time it takes each day to implement the treatment plan
	2. Minimize the time it takes to obtain authorizations (e.g., provide justification for procedures, medications, equipment, etc.)	2. Minimize the length of time the treatment plan must be conducted (e.g., days, weeks, etc.)
	3. Minimize the likelihood that the care will be delayed in order to get the authorization	3. Minimize the time it takes to select a treatment plan from the options (e.g., one that is consistent with personal preferences, one that can be remembered, one that is not too complicated, etc.)
	4. Minimize the frequency with which a provider must seek approval for a standard course of treatment	4. Increase the likelihood that the treatment plan can be executed while away from home (e.g., supplies are portable, daily activities can be maintained, etc.)
	5. Minimize the likelihood of being limited in treatment options because of health plan restrictions	5. Increase the likelihood of sticking to the treatment plan
	6. Minimize the likelihood of selecting a treatment plan that the patient cannot afford	6. Minimize the number of treatment plans that must be tried before the condition is resolved

The use of outcome statements is highly effective when you're trying to identify an opportunity in a mature market where there is already a high level of satisfaction and the innovation team is looking for a hidden opportunity—areas that are difficult to see at the broader level. In most social scenarios, this is not the case; the dissatisfaction present in most social scenarios is such that it can easily be understood at the job step level. In such cases, it is more effective and just as reliable to use detailed job steps and forgo the outcome statements. Let's look at the difference in the two levels of need statements. If one uses *outcome statements* as the form of needs as shown in Table 1-6, then the job step would be "establish a treatment plan" and would not be rated as a need, but is primarily, a category placeholder. In this case, the *outcomes* are the needs that are used for rating and purposes of innovation.

Contrast this approach with that of using the job steps as the need statements where, in the table, the execute category would have been broken down further to include *all* the job steps involved in the execution of the focal job. In this case, "establish a treatment plan" would have the following job steps (with no outcomes):

1. Identify lifestyle issues that may prevent the patient from being able to adhere to a treatment plan, e.g., travel, work, etc.

2. Determine the optimal treatment plan for the patient based on his or her condition and lifestyle

3. Determine whether the treatment plan is affordable for the patient, e.g., covered on his or her plan, lowest out-of-pocket costs, etc.

4. Obtain the necessary authorizations for the treatment plan

5. Gain approval from the patient to proceed with the plan, e.g., discuss options, answer questions, etc.

6. Order the agreed-upon treatment plan, e.g., write prescriptions, order tests, prepare instructions for the patient, etc.

As is evident in the example, the statements for the execute step are sufficient for understanding the needs of the physician. Additionally, they are more concise, easier to read and comprehend, and are less likely to confuse the patient.[16]

After the statements have been rated for importance and satisfaction, then we can explore these metrics in the contextual interviews that are conducted on the high opportunities. Thus we gather the necessary detail only on those items that have been deemed a good opportunity (very important and not well satisfied). Using the example above, assume that one of the high opportunities turns out to be "obtain the necessary authorizations for the treatment plan." During the contextual interviews, this statement would be explored further along a number of dimensions, including speed stability and output (this process is discussed more fully in Chapter 3).

In social innovation, the job steps are sufficient as needs and forgoing the outcome level to be skipped for three key reasons. First, the job steps are simpler and more straightforward, thus easier for the respondent to answer and the innovation team members to understand consistently. Second, this process produces a much shorter set of needs resulting in a more abbreviated questionnaire which yields lower costs and likely better response rates. And finally, the brevity of the statements allows for better comparison among the members of the ecosystem, thus enabling the innovation team to better discover the synergies and conflicts.

Related Jobs

The related jobs are the additional goals, activities, or tasks that the member is trying to achieve *while executing the focal job*. Thus, in the case of physicians, we would be looking at what other jobs they are trying to do while treating a patient. In studies we've found that physicians are also trying to:

Engage in Professional Development

1. Stay on top of new techniques, best practices, cutting-edge treatments

2. Influence the development of medical technology, e.g., new products, new medications, etc.

3. Improve professional skills through exposure to various types of situations

Handle Automation and Technical Advances

4. Set up the practice on electronic medical records

5. Convert from paper claims to electronic claims

6. Train the staff on new automation solutions, such as electronic patient records, claims, and so forth

Manage the Business

7. Hire staff, e.g., front office staff, physician assistants, technicians, etc.

8. Determine the optimal appointment scheduling flow, e.g., for example, how many patients can be seen in a day, how long each visit should be, etc.

9. Ensure that the amount of time scheduled for an appointment is sufficient to complete an evaluation and diagnosis

Work with Patients to Improve Their Health and Well-Being

10. Educate patients on what is needed to improve their health, e.g., lifestyle changes, preventative measures, diet, exercise, etc.

11. Respond to patient inquiries, e.g., phone calls, e-mails, etc.

12. Overcome patient misinformation, e.g., provide patients with the correct information, inform patients of what is really necessary, etc.

This is only a subset of the jobs identified for physicians, but it provides an example of the types of related jobs we expect to see.

While we often see innovation take place when the focal job is improved, a successful innovation strategy will also explore other jobs that can be addressed as well. If we take the iPhone as an example in the product space, it is clear that Apple succeeded in part by creating a great communication experience; however, the real value of the iPhone is all of the related jobs it helps people get done while on the go.

Thus, understanding related jobs can expand the perspective and possibilities for innovation of the ecosystem members. For instance, when patients are trying to manage chronic diabetes, one key job they are trying to get done is to, "determine their blood glucose level at any given time." There are many different types of blood glucose meters that can address this job. Some meters have made the job of retrieving a blood sample easier by having smaller lancets, having the equipment to retrieve the blood all contained in a cartridge, and requiring smaller amounts of blood to read the sugar levels. However, there is also a software program introduced by OneTouch that helps patients get more jobs done while managing their blood glucose level. This software exports data from the device and provides reports which enable the user to:

- Identify trends in the blood glucose levels
- Identify meal-related patterns
- Identify high and low blood glucose levels

All these related jobs are also very important to people trying to manage their diabetic condition. This example shows how adding features to a product, service, or program can help people get other jobs done in addition to the focal job.

Let's look at an innovative program in the education sector. A top issue for students and parents of low-income working families is the need for after-school care. While the primary job that after-school programs typically address is to "provide a safe environment for the child to be between school and the time parents get off work," a group called Heart House has addressed many more jobs than simply providing child care at a low rate. It addresses several other jobs that are important to both parents and students, including:

- Ensure children complete their homework
- Assist children in learning concepts that they are struggling with

- Help children gain an understanding of what behaviors are needed to become a responsible citizen
- Help children work through life stresses
- Improve childrens' reading capability
- Help children succeed in their school work
- Prevent children from getting involved in gangs

A comprehensive picture of the issues within the social scenario requires an understanding of both the focal job and the related jobs of all of the members, providing a means of innovation from a number of angles.

Let's look at an example of focal jobs and related jobs within an ecosystem. An organization hoping to provide nutrition to children in developing countries may frame the innovation strategy as shown in Table 1-7.

Keep in mind that whether or not we label a job "focal" or "related" depends entirely on the perspective of our investigation. A job that we call "related" in one investigation might be called "focal" in another. For example, using the health care system case above, if the innovation program is being driven by members within a state government agency responsible for managing the costs of health care for the uninsured, the focal job is likely to change. Let's assume that one of the main drivers of cost is the treatment of complications from chronic conditions such as diabetes, high blood pressure, and so on. In this case, the patient's related job of managing a chronic condition to prevent it from getting worse may become the focal job of the new innovation program.

Emotional and Social Jobs

Next we round out the needs with two types of jobs that often play a significant role in social innovation scenarios: the emotional and social jobs. These are the jobs that the ecosystem members want to get done with respect to how they want to feel (emotional jobs) and how they want to be perceived (social jobs). To get to the emotional jobs, we might ask

Table 1-7 Social Scenario: Provide Nutrition to Children

Social Scenario	Provide Nutrition to Children		
Context	In Areas Affected by Severe Malnutrition		
Ecosystem Members	Beneficiary Parents	Executor Clinic Workers	Overseer/Third Party Aid Agencies
Want to accomplish this main thing	**Focal job** Provide food to the family	**Focal job** Distribute food to families	**Focal job** Manage the logistics of food provision
They must also	**Related jobs** Determine what a healthy weight is for the child Identify dietary changes needed to improve the child's health Identify signs of illness in the child Ensure that the child's siblings maintain a healthy weight	**Related jobs** Educate parents on proper nutrition for the child Determine if the child has improved since the last visit	**Related jobs** Identify geographic areas where people are suffering from malnutrition

the respondents not only how they want to feel as they are involved in the social scenario, but also what their aspirations are, how they define their values, what inspires them to participate. On the social side, we will ask how they want to be perceived, as well as how they engage with others, how they want to participate or interact with others, and so forth. The list contains a set of emotional and social needs of physicians while they are practicing medicine (this is only a subset of the needs identified).

1. Feel appreciated by patients
2. Feel respected by others—peers, administration, patients, and so on
3. Feel in control—avoid having others second-guess your decisions
4. Feel like a vital part of the community
5. Feel like you are making a positive difference in your community

6. Enjoy people, e.g., the patients, staff members, colleagues, etc.

7. Help sick people to regain their health and vitality

8. Restore function to those who are injured

9. Be perceived as competent

10. Be perceived as an expert in one's chosen field

11. Be the "go to" person for difficult issues

12. Enjoy the challenge of intellectual debate with colleagues

Constraints

The last, but critical, element in mapping the social scenario is to identify the constraints that exist. There are two types of constraints we need to understand: environmental constraints and human constraints. Environmental constraints are the physical, technical, political, or geographic issues that must be overcome when we attempt to solve the problem. We covered some environmental constraints in the case of Plumpy'nut: lack of refrigeration, lack of potable water, very little storage space, and so on.

Human constraints are the emotional, mental, social, and cultural issues that must be taken into consideration. In the case of getting aid to disaster victims, for example, it is important to take into account the fact that the victims are likely to be confused, disoriented, and frightened. In the health care scenario, we face issues such as patients who simply refuse to change their lifestyle, even though failure to do so means that they will continue to get sick. In schools, we have stressed teachers who are fearful of violence because of the increasing number of shootings on school campuses. These human constraints can affect how the members of the ecosystem respond to different solutions and how motivated they are to give a solution a chance.

In our study of the health care scenario, physicians identified the following sample of constraints as critical to the success of our health care system:

Human Constraints

1. Patients do not take responsibility for their wellness and will not make lifestyle changes
2. Physicians are overtired and often work with little sleep
3. Lack of many specialties overload the ones that are there
4. Physicians are frustrated by the fact that convenient methods of communication, such as e-mail, phone, and the like, are not reimbursed and therefore are not used as much as they could be

Environmental Constraints

1. The cost of training new providers is so high that physicians must make large salaries to pay back student loans
2. There is an anticipated shortage of key types of providers
3. Misaligned incentives cause overutilization or underutilization of medical services
4. The industry does not financially reward prevention and wellness

Consider environmental constraints in education: one of the largest environmental constraints that students in the inner cities face is a lack of access to technology that can help them with their schoolwork. Knowing that this problem exists, city governments can work with nonprofit groups, corporations, and government entities (public libraries, for example) to make the technology available to all students. The Race to the Top Fund, for example, is a $4.3 billion fund that uses competition to spur innovation. States compete for money by developing innovative programs to improve their educational systems. Money is distributed based on the performance and outcomes of the ideas at work. Then there's the New Schools Venture Fund, which was created in 1998 by Kim Smith (a social entrepreneur) and venture capitalists John Doerr and Brook Byers. It created a community of education reformers who were in the field every day fighting to dramatically improve educational outcomes for underserved students in grades K–12. These are just a few examples of organizations banding together to try to overcome the constraints that prevent satisfaction among the ecosystem members. What is critical, however, is a full understanding of all of the constraints and a prioritization of those constraints in order to make these programs more effective.

Needs of Third Party and Overseers/Constraints

The third-party members of the ecosystem are vital for understanding the links and connections within the ecosystem, the jobs that connect the members to each other, and the criteria they use to assess one product, service, or program over another. Thus, to begin to identify needs of the third-party members, it is important to first diagram the relationships between this member and the other members of the ecosystem. For example, in the education scenario, the following table of members exists:

Job Executor	Job Beneficiary	Third Party/Overseers
Teacher	Student	Superintendents
Parent	Parent	Principals
Tutors	Businesses Who	Technology Directors
	Will Hire Students	Curriculum Directors
	Colleges Who Will	Library/Media Staff
	Receive Them	Dept of Education Personnel

So, in the case of Technology Directors, the diagram would have the Technology Director in the middle, with connections to each of the other third parties, to the job executors and possibly to the job beneficiaries (some overseers have direct ties to beneficiaries and some do not). For each of these connections, it is important to identify the jobs the Technology Director is trying to get done. For instance, in the connection with the Teachers, the Technology Director is trying to "troubleshoot problems with equipment", "order new equipment'" "dispose of broken equipment," and the like. In the connection with the Superintendent, the Director has several different jobs pertaining to the recommendation of new technologies, evaluation of new technologies, and so forth.

With many of the overseers, they have a set of criteria they are using to evaluate a product, service, or program. These criteria are important for purposes of innovation as they will provide needed guidance in the design of the innovative solution. A national organization that makes purchase decisions on what kind of food aid to purchase, they may have some of the following criteria:

Structural Considerations

1. Ensure the food is shelf stable, e.g., does not require refrigeration
2. Reduce the amount of spoilage of the food aid, e.g., during transportation, during long periods of storage, etc.

Cost Considerations

3. Individual organizations must be able to purchase the food aid
4. Reduce the fluctuation of food aid pricing year over year
5. Ensure food aid can be purchased by those with very low income levels

Organizational Considerations

6. Help organizations to determine how much of the food supply is needed

Distribution Considerations

7. Be able to distribute the food aid to remote areas
8. Reduce the cost of distribution to remote areas

Functional Considerations

9. Ensure the food can be prepared without water
10. Allow the food to be eaten without help from the parent, e.g., child friendly

There are other considerations to think about as well including:

- Purchase process
- Service after purchase

- Raw materials
- Sustainability
- Usability

We have a sample moderator guide with instructions at our website (www.TheSocialInnovationImperative.com).

Stability Over Time

With all the types of needs we've discussed, it is important to emphasize that these needs must be *free from solutions, unambiguous, and worded precisely and concisely.* The reason behind this precision is that these needs, when properly stated, are constant and stable over time. The needs that exist today are the same needs that people had two decades ago. Physicians have always been trying to treat a patient's illness, teachers have been trying to ensure that students understand new content, first responders have been trying to obtain the vitals of an accident victim, and so forth.

The job steps are also stable over time to the degree that the execution of the job has not dramatically changed. For instance, because we are not using technology or products or solutions in the statements, the job step of "confirm the treatment plan with the patient" is one that is likely to have been conducted 50 years ago and will still be executed 50 years from now. Some job steps may have been added or removed as the job itself changes. For example, before there was insurance to pay for standard medical treatment, there was likely not a step for "obtain authorization for the treatment plan." It also may become obsolete with the disruption of the insurance and health care industries.

Thus, we consider the need statements, whether jobs, related jobs, job steps, outcomes or constraints, to be constant and stable over time. What does change is the importance and satisfaction of the needs, driving changes in the priority of the needs for purposes of innovation. Thus it is vital to update the importance and satisfaction of these needs on a regular basis in order to identify new trends or to determine the effectiveness of new programs and services.[17]

In Summary

Understanding the needs and constraints of the ecosystem members is one of the most critical parts of the innovation process and what makes the Social Impact Framework outlined in this book so powerful. Techniques used in the corporate world have generated great success and resulted in the launch of numerous new products and services. Our goal is to see this same level of success when the process is applied to the social issues that plague our world. The jobs-to-be-done method of understanding needs has become well known and has the unique advantage of being able to identify market need in areas where there is no product and thus no market to measure. In cases where organizations struggle to find what the people want, using jobs can help identify exactly where the need is and where value can be created.

Chapter | 2

UNDERSTAND AND
PRIORITIZE THE NEEDS

A s we discuss in Chapter 1, social challenges can be explored through the lenses of the jobs-to-be-done of the ecosystem members. More specifically, identifying synergistic jobs and conflicting jobs is vital in order to balance members' job satisfaction through innovative programs, products, and services. This chapter focuses on how to capture the members' needs (through qualitative research techniques) and prioritize these needs (through quantitative research methods).

A common mistake made in innovation programs is to focus on only one or two members of the ecosystem. Such a myopic view leads to solutions that often fail to satisfy others in the ecosystem whose needs were not considered, yet are vital for adoption of the innovation. For example, trying to devise innovations in the health-care space without considering vital ecosystem members such as employers and insurance companies could leave serious gaps in the knowledge of the purchase and implementation requirements of these parties that control such a significant part of the health care ecosystem.

Designing the Social Innovation Blueprint

The *social innovation blueprint* is a worksheet for visually mapping the innovation challenge. It brings clarity to the key questions: who, what, when, and how. Who is involved? What do they need to accomplish? Does what they want to accomplish change over time (when)? How must the job be executed? (That is, what constraints must be overcome?) This section provides step-by-step instructions on how to create this blueprint.

Who?

The first step is to dissect the social challenge in order to identify all the critical members involved and how they interact. To begin, create a simple Venn diagram (a series of overlapping circles which show the relation of finite sets, as shown in Figure 2-1) that illustrates how the members come in contact with one another. This visualization helps teams of people come to agreement on some of the basics, providing focus and clarity for the starting point. The goal is to untangle the elements of the problem and put the pieces in a framework that will allow the problem to be worked on in a methodical and systematic way.

Start by identifying the core job executor and the core job beneficiary since they are the primary targets of the innovation. In the Plumpy'nut case the primary beneficiary is the child, and the primary executor is the clinic worker. Next determine if there are secondary beneficiaries and executors. In the Plumpy'nut case there is a secondary beneficiary—the parents. The needs of the overseers and third parties will also have to be considered, such as aid agencies and government agencies.

Keep in mind that when studying synergy within the ecosystem, it is important to explore the same *experience* from the vantage point of different members of the ecosystem. For example, when identifying synergies in the school system, we would want to study the job "teach new content" and "learn new content," which will allow us to see the entire scenario from different perspectives. In the health-care system, the physician's job of

Figure 2-1   Experience of Health Care Delivery

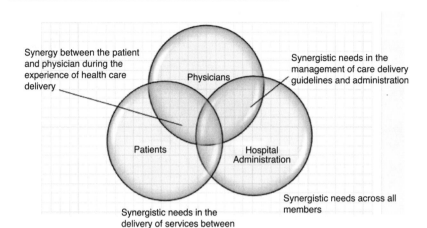

Synergy between the patient and physician during the experience of health care delivery

Synergistic needs in the management of care delivery guidelines and administration

Physicians

Patients

Hospital Administration

Synergistic needs in the delivery of services between the hospital and patient

Synergistic needs across all members

"treat a patient" and the patient's job of "obtain care" are both part of the experience of the delivery of care to patients. Both of these focal jobs should be studied to find the synergies in how the job is executed and received.

As we consider beneficiaries it is important to note that the beneficiary might not always be an active *participant* in the scenario. For instance, gang members who are targeted for reform are the beneficiary, yet they are not likely to be a "voluntary" participant. An abused child might be the beneficiary of actions to stop domestic violence, but they will be a difficult audience to speak with. Other instances where the beneficiary might not be able to "speak" for themselves include those with cognitive impairment, those with Alzheimer's or stroke patients, and so on. These beneficiaries are important to recognize as the blueprint is being established and it is vital to determine how information will be obtained by these beneficiaries. In these cases, ethnographic research can be used to watch their behaviors to document the jobs and/or interview relatives, caregivers, or other parties who are familiar with the situation.

The executors and beneficiaries are listed in the Social Innovation Blueprint in their appropriate sections, as shown in Figure 2-2. Third parties are listed with the job executors.

Figure 2-2 Innovation Blueprint

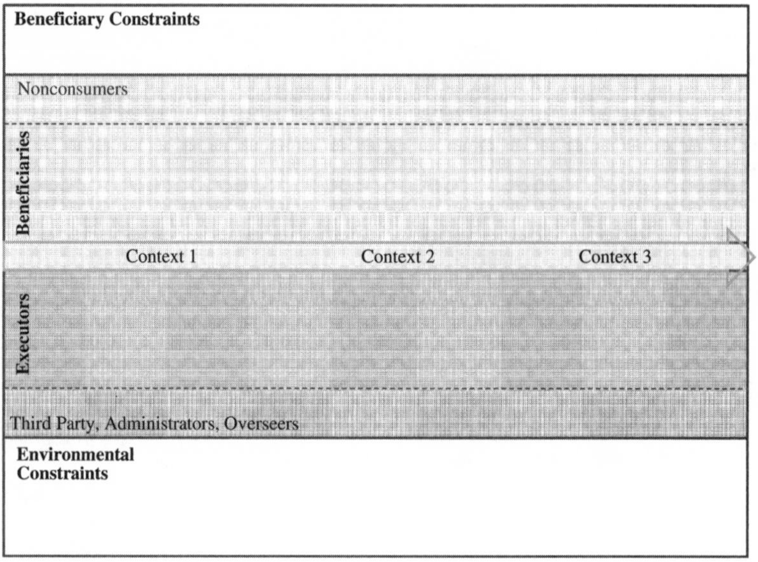

What?

Next determine the *primary* job-to-be-done for each of the members. What is the primary role each plays in the scenario? What is each of them trying to accomplish within the scenario? While the job beneficiaries and job executors are trying to get multiple jobs done, identify the focal job for the blueprint. Examples of focal jobs in the Plumpy'nut case are listed below:

- The child is trying to recover from the effects of malnutrition.
- The parent is trying to get the child care for the effects of malnutrition.
- Clinic workers are trying to provide the child with care.
- Aid workers are trying to distribute supplies to clinics.
- Government agencies are trying to determine how to distribute aid in an orderly fashion.

In the first section of the Social Innovation Blueprint, fill in each of the members of the ecosystem in the first column in addition to the focal job each is trying to get done.

When? Or Where?

Next, determine what contextual changes should be studied. In oth
how does the scenario change over time or in different places? What a. ome
of the key points at which the job executors and/or the jobs of the existing members will change? These context changes can be based on a circumstance, event,
time frame, or something else. The goal is to understand how the members
change, as well as how the jobs and needs change when the context changes.
It is important to study both the normal state of affairs and some "crisis" or
complex state, since during such states, the jobs' priorities may change, and it
is important to know these new priorities and understand the trade-offs. In the
Plumpy'nut example, the status quo job is providing nutrition for the child. But
when the child comes to the clinic severely malnourished, the job changes:
the crisis jobs include treating the child for malnutrition, preventing complications from malnutrition, and educating the parents on how to continue treatment at home. Then there are the postcrisis jobs, when the child returns home
after treatment. The parents are the job executors at that stage, and their jobs
include ensuring that the child continues to get the necessary nutrition, making changes in the way the household food distribution takes place, and so on.

In the Social Innovation Blueprint, map out the changes in the context
and then identify job executors and beneficiaries for each of the new contexts. If the same members are present, document what their new focal job
is in the different context as it is likely to change with the context.

How?

In all social scenarios, there are constraints that must be overcome if a successful solution is to be executed. The beneficiary is generally constrained
primarily by what we call human constraints—the personal, social, and
psychological constraints that affect his or her ability to get the jobs done.
The executor is constrained primarily by environmental constraints, which
include the technical, political, social, and physical problems that can prevent
the job executors from effectively getting the job done. For purposes of the
blueprint, we want to identify the initial constraints that we believe we will

encounter in the social scenario. These must of course be validated and will be further explored during the qualitative interviews. Generally, however, the team involved in framing the scenario will have a pretty good idea of the constraints that will be encountered which can be entered in the blueprint.

In the Plumpy'nut example, constraints that prevented beneficiaries and executors from getting jobs done included cultural issues surrounding the distribution of food in the household, lack of refrigeration, lack of storage space for food, and lack of clean water for cooking, among other things. Plumpy'nut overcame those constraints in a truly elegant way: it requires no refrigeration, the packets it comes in are lightweight and very portable, and it is presented as a food that is specifically for children, so it bypasses the food distribution problems inherent in the family meal.

Let's look at an example in education for low-income areas. In this case, there are two job executors—teachers and parents—and the job beneficiaries are the students and secondarily the department of education in the various states. The Social Innovation Blueprint example shown in Figure 2-3 illustrates the context of taking a standardized test, such

Figure 2-3 Innovation Blueprint for Education in Low-Income Areas

Challenge: Education in low-income areas

	Human constraints		
	Multitasking multimedia-raised children are easily bored	Children do not see the connection of these tests to their life	
Beneficiaries	**Student** Job: learn content	**Student** Job: complete the standardized test — State education departments: assess how well students are learning material; compare schools' competence	**Student** Job: Use scores to ???
	Status Quo	Crisis period standardized Tests?	Postcrisis
Executors	**Teachers** Job: communicate content to students; ensure student learns material — Parents: provide encouragement to the student, provide help/support to the student	**Teachers** Job: conduct the standardized test — Parents: ensure students are prepared for the test	**Teachers** Job: justify/support scores, adjust lesson plans to improve
		Often either both parents working or it is a single-parent household	
	Environmental Constraints		Teachers are often overloaded—too many students per teacher

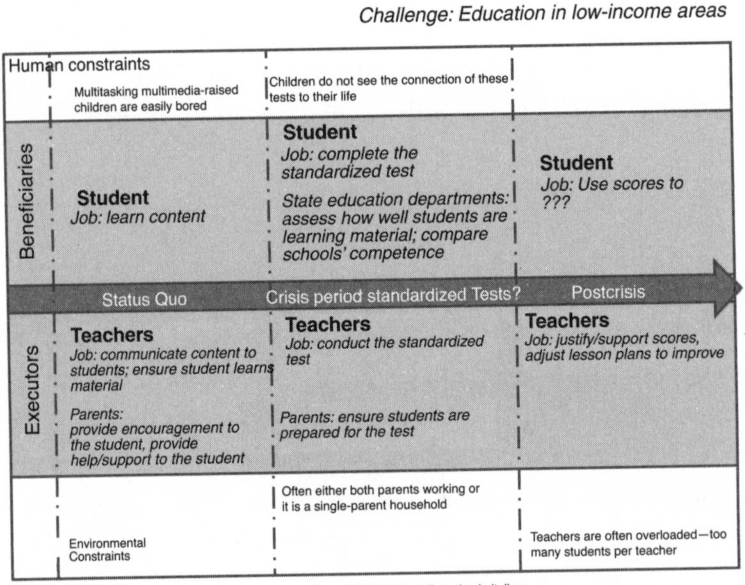

Bold indicates primary beneficiary and primary executor   Job is listed for all parties in italics

as those required under the auspices of the Elementary and Secondary Education Act (the "No Child Left Behind" Act).

Let's look at another example. In the case of health care, the jobs that people want to get done on a daily basis include taking care of bodily needs, maintaining a desired level of physical fitness, and so on (see Figure 2-4). When people are sick or injured, however, their primary job becomes to seek care while they are sick, which includes such subordinate jobs as determining whether the illness will go away without help, determining what type of doctor should be seen, and so forth. After seeking help for the illness, people may go back to their day-to-day routine, or their job priorities may change and they may seek to eliminate a bad habit related to the illness (e.g., not getting enough sleep, drinking too much coffee, etc.), so that they can avoid having to tackle the job of seeking care when they get sick again.

These examples illustrate how the blueprint is able to visually present the scenario as it is executed over time. This simple tool can provide the

Figure 2-4 Innovation Blueprint—
Health Care Delivery for the Uninsured

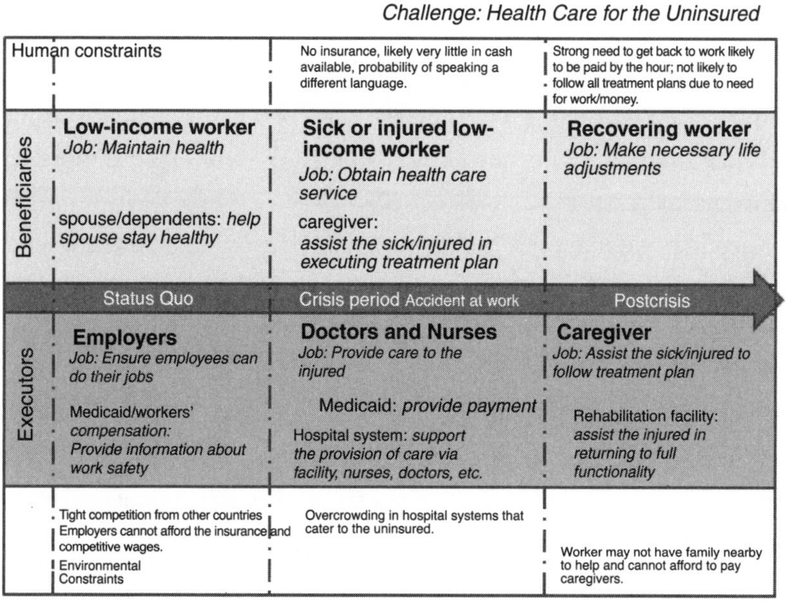

Challenge: Health Care for the Uninsured

	Status Quo	Crisis period *Accident at work*	Postcrisis
Human constraints	No insurance, likely very little in cash available, probability of speaking a different language.	Strong need to get back to work likely to be paid by the hour; not likely to follow all treatment plans due to need for work/money.
Beneficiaries	**Low-income worker** *Job: Maintain health* spouse/dependents: *help spouse stay healthy*	**Sick or injured low-income worker** *Job: Obtain health care service* caregiver: *assist the sick/injured in executing treatment plan*	**Recovering worker** *Job: Make necessary life adjustments*
Executors	**Employers** *Job: Ensure employees can do their jobs* Medicaid/workers' compensation: *Provide information about work safety*	**Doctors and Nurses** *Job: Provide care to the injured* Medicaid: *provide payment* Hospital system: *support the provision of care via facility, nurses, doctors, etc.*	**Caregiver** *Job: Assist the sick/injured to follow treatment plan* Rehabilitation facility: *assist the injured in returning to full functionality*
	Tight competition from other countries Employers cannot afford the insurance and competitive wages. Environmental Constraints	Overcrowding in hospital systems that cater to the uninsured.	Worker may not have family nearby to help and cannot afford to pay caregivers.

Bold indicates primary beneficiary and primary executor   *Jobs is in italics* necessary framework for exploring the situation and can ensure common understanding among members of the innovation team.

Framing the Initiative

Once the blueprint has been worked out, the results can be put into the Innovation Framing Guide, a tool that will help document the decisions made regarding the innovation initiative. This step is necessary since in most cases innovation teams will not be able to tackle everything present in the blueprint. Thus, decisions must be made as to which context(s) to work with, which executors, beneficiaries, and third parties. For instance, once the various contexts have been mapped out, which members of the ecosystem will be studied? Which contexts? Which jobs? This guide is valuable in making sure the innovation team focused on the same problem scope, the members, the jobs, and so on.

The framing guide has some additional questions to be filled out at this point. The current platforms available to the ecosystem members to get the job done should be listed, as well as the current ways the members get the job done, technology considerations, and what an ideal solution would look like for the ecosystem member. These additional insights can be added to over time and will help the team as the project progresses. It is important that there is an understanding of what is currently available and how the scenario is currently working. We will explore more about platforms later.

Keep in mind that since the members of the ecosystem have not yet been interviewed, there are many areas that may be left blank. This is not a problem; instead it serves to provide insight for areas to explore in the interviews with the ecosystem members. The goal is to think through what you already know about the scenario and the members, jot down some examples, and highlight areas where the knowledge is very weak. This will help in formulating the screening for the interviews by identifying areas of special focus and questions that need to be answered by the stakeholders. Table 2-1 provides a template of the Innovation Framing Guide. This tool is also available for download at our website www.TheSocialInnovationImperative.com.

Table 2-1 The Innovation Framing Guide

Social Scenario			
Context			
Ecosystem Members	**Beneficiary**	**Executor**	**Overseer**
Want to accomplish this main thing . . .	Focal job	Focal job	Focal job
Ecosystem members measure success in getting that job done in terms of . . .	Jobs steps of the focal job	Job steps of the focal job	Job steps of the focal job
They also want to get these related jobs done . . .	Jobs related to beneficiary's focal job	Jobs related to executor's focal job	Jobs related to overseer's focal job
These human constraints (social, cultural, psychological, biological, etc.) are likely to exist	What prevents the beneficiary from being satisfied?	What prevents the executor from being able to deliver?	What human issues must the overseer consider when trying to improve satisfaction?
These physical/ environmental constraints (physical environment, government or institutions, technology, etc.) are likely to exist	What physical constraints that prevent the beneficiary to get the job done	What physical constraints prevent the executor from being able to satisfy the beneficiary	What physical constraints exist for the third party and overseers
The platform that exists today includes . . .	Platform description (systems, subsystems, people, processes, and technology)		
Current ways of tackling core and related jobs include . . .	Competitive alternatives including nontraditional solutions that can be used to get the job done		
Technology considerations include . . .	Examples of technologies that exist or that have been recently introduced to support this job(s)		
This initiative will create what "ideal solution"?	If this initiative is successful, what will be different? How will the scenario be affected? What kind of solutions will be created?		

With the health-care example mapped in the Social Innovation Blueprint with uninsured workers in Figure 2-4, decisions are ready to be made as to which members, which jobs, and which context to focus on. These decisions are then transferred to the framing guide. The Innovation Framing Guide shown in Table 2-2 is filled in based on the "Crisis" context, with the injured worker as the beneficiary and the medical staff and hospital system as the executors.

Table 2-2 Health Care for the Uninsured

Social scenario	Health care for the uninsured		
Context	Accident at work		
Ecosystem Members	Beneficiary Patient	Executor Physicians, Nurses, Physician Assistants	Overseer/Third Party Hospital Administration
Want to accomplish this main thing	**Focal job** Obtain health-care services for the injury	**Focal job** Treat the injury	**Focal job** Create an environment that optimizes the treatment of patients
They must also	**Related jobs** Obtain assessment of the injury Determine how bills will be paid Determine how long the recovery time will be File workers' compensation claim	**Related jobs** Determine what other conditions the patient has Determine whether the patient is able to return to work	**Related jobs** Determine whether workers' compensation is to pay for the claim Determine patient satisfaction
Human/ environmental constraints	Frustration/anger, worry about how bills will be paid, worry about losing job	Communication issue if the worker does not speak English well Extremely busy ER environment	Overcrowding in ER and hospitals that serve the indigent Unable to recoup costs of treating the patient
The platform that exists today		Hospital system Workers' compensation Medicaid	
Current ways of tackling core and related jobs	Go to ER since there is no insurance coverage	Care for any who come to the ER; first come first served plus triage	Social services can assist with the patient's questions
Technology considerations	Centers that focus on urgent care needs that are nonemergent/less expensive, would free up ER for true emergencies, offload some of the Medicaid burden to less expensive facilities, etc.		

What is the ideal solution?	Injury gets treated for a fee that is affordable; find a way to keep working through the injury period	Reduce the patient flow to a reasonable level; ensure that all patients can be treated in a reasonable time frame	Get rewarded for the care provided
		Have support to communicate with patients who do not speak English well	

Capturing Need Statements

As discussed in Chapter 1, the needs of the ecosystem members must be learned from the ecosystem members themselves. This section covers how to capture the job steps, related jobs, the emotional jobs, the constraints, and the inputs of the third-party members.

Methods of Capturing Needs

Gathering information on the needs of the ecosystem members can be accomplished in a number of ways. Any of the traditional types of qualitative research techniques can be used to garner the information, including focus groups, in-depth interviews, contextual interviews (sometimes referred to as ethnographic interviews), day-in-the-life diaries filled out by the individuals, and so on. The goal is to capture needs in the way that is the most comfortable for the respondent and that ensures a complete understanding of the needs.

Focus groups are beneficial when the goal is to create dialogue, to gain inputs from a large group of people in a shorter period of time, or when there are several stakeholders sponsoring the initiative that want to view the interviews. Focus groups are also beneficial when trying to understand what makes a product or service complex or difficult. For instance, when trying to create a disruptive platform for people to use for monitoring their health statistics, it might be beneficial to have the existing monitoring devices in the room for them to interact with and then describe what makes working with them difficult or confusing. Keep in mind though that we are not trying to capture the properties of the solutions—we are trying to understand the customers' needs. For instance, if the users indicate that the health monitoring devices are difficult to use, we'd want to find out specifically what skills they feel they lack in order to use the device. It may be that the need of the respondent is to avoid making a mistake when taking vital signs, or to be able to understand the meaning of the output of the vital signs; for example, is it good or bad? What should be done about it? Or how does a customer with poor dexterity (e.g., arthritic hands) take vital signs? The tools used in the focus group function simply as props to get the respondents to express what they are trying to get done; one must be very careful not to turn the group into a "concept review" of the products that are being used as props.

Focus groups are also beneficial when dealing with large groups of citizens, patients, teachers, or any other group in which a great deal of diversity is desired.

Phone interviews are very effective tools when trying to dive deep into an area of interest. These are beneficial for cases when the ecosystem members are difficult to reach or difficult to schedule in a group setting such as field workers, executives, and physicians. Phone interviews are generally about an hour long and are usually recorded for future reference.

When possible, in-person interviews are preferred. They make it possible for the interviewer to see and understand the environment in which the job is being executed, and to watch the member perform the job to see what struggles are encountered. In-person interviews are two hours long and are often supported by a videographer to capture visual information about the ecosystem member, the physical environment, and any work-arounds the member does when he or she can't get the job done directly. These visuals become very important in understanding the context and the constraints, as well as providing video clips to support the job steps documented.[1]

Tim Brown, CEO of Ideo, created a technique called Design Thinking. In his article on *Design Thinking for Social Innovation*, he emphasizes the importance of in-person interviews and for the innovation team to "go out into the world and observe the actual experiences of [members] as they improvise their way through their daily lives." He emphasizes the need to work with local partners to establish credibility and so that they can make the necessary introductions to the community in which the members reside. "Through 'homestays' and shadowing locals at their jobs and in their homes, design thinkers become embedded in the lives of the people they are designing for."[2]

Generally, two to three phone interviews plus four to six in-person interviews are needed to ensure that all the needs and constraints have been captured. The key is to get a highly diverse group to ensure that all of the possible types of ecosystem members have been covered. For instance, if one of the member groups is hospital administrators, small hospitals as well as large, urban as well as rural, teaching as well as nonteaching facilities should be explored. Focus on getting the type of diversity that exists in the market to ensure that the job is documented in a way that accommodates all the variations in the execution of the job. A sample plan for the qualitative interviews may look like the information in Table 2-3

The interview guide for these programs should be structured before the interviews begin. My colleagues and I created a worksheet that provides a nice guide as well as a place to document the responses. We use a standard Word document with a list of the universal job-step questions, prompts for related jobs, and prompts for emotional jobs in a sidebar on the right-hand side. This gives us the prompts we need as we go through the interview as well as plenty of room to document the answers. A small sample of this type of guide is shown in Figure 2-5.

Table 2-3 Qualitative Interview Plan for Health Care Study

Patients	25–34	35–55	56+	Total
No chronic conditions	2-1 phone, 1 onsite	2-1 phone, 1 onsite	2 onsite	6
Have chronic condition	1 onsite	1 onsite	1 onsite	3
Total	3	3	3	9

Figure 2-5 Moderator guide for qualitative interviews.

Ecosystem Member Name

Lead-ins for the job statment:
How important is it to you that you are able to . . .
How satisfied are you with your ability to . . .

When does the job actually begin from your perspective?

Define

1. Job step
2. Job step
3. Job step

Define (Path, select, determine):
What must be determined up front, before you start the job, to ensure that the job will be successful?

What must be defined? What must be planned?

Locate

What else? Why?

4. Job step
5. Job step
6. Job step

Locate (gather, access, retrieve)

What must be retrieved or gathered before you can begin the job?

Probe: information? Physical objects? People?

Prepare

7. Job step
8. Job step

What can become a problem if you don't have it when you start the job?

This or any tool that provides prompts for remembering the goals and keeping on track is important since it is likely that the respondents will not stay on track and will start talking about solutions that they find appealing or take you on a tangent about policy issues, or something similar. The guide ensures that the conversation will stay on track, and it provides a means of pulling the respondents back on track if they begin to wander.

Capturing Jobs

Once the focal job is identified and framed, it is time to begin the interviews to capture the job steps, related jobs, and emotional jobs. To develop the initial job map within the categories of the universal job steps and to gain an understanding of the overall ecosystem, it is recommended that phone interviews be used. These interviews will provide an overall foundation for the next set of interviews. These first interviews can also be used to explore contexts, current solutions being used, platforms, and other vital information that can be useful in the more detailed interviews.

During the interview, confirm the job map and then begin to fill in the detailed job steps. Allot time to ensure that the emotional jobs, related jobs, and constraints are also addressed.

Keep in mind that it is alright that the members we speak with may not execute the job in the same order or with exactly the same process. Because the job steps are documented based on what the respondents are trying to accomplish, not what they are actually doing, it will not be an issue if they execute the process in a slightly different order or if they execute a step that another member doesn't. The question is why are they executing that step? If they are actually accomplishing a new job that the others don't do, then we will capture it for the one circumstance that has it. The quantitative analysis will let us know how many ecosystem members find that job step to be important. The qualitative analysis is not designed to gain consensus or to quantify how many members execute the job in a certain way. The goal of the qualitative analysis is to uncover all possible job steps so that they can be rated for importance and satisfaction in the quantitative study.

The questioning for determining the job steps is documented in detail in the *Harvard Business Review* article developed by Lance Bettencourt, titled "The Customer-Centered Innovation Map." For those who plan to conduct these ecosystem member interviews themselves, this is a must read.[3] In order to learn more about capturing emotional jobs, related jobs, and social jobs, please refer to *What Customers Want* by Tony Ulwick. Chapter 2 of Ulwick's book provides excellent detail as to how to capture the needs.[4]

Capturing Constraints

The constraints that exist within the social scenario can be numerous and will relate to each member of the ecosystem we are working with. Therefore, in the education scenario, we would capture constraints from teachers, students, parents, school administrators, and so forth. There may be some overlap, especially in the environmental constraints since many of these members are working in the same environment, and this is fine. We will still want to document that the constraints exist in each of the members' areas in which they are found.

To capture the human constraints, we need to probe the behavioral, psychological, social, or cultural issues that affect the ecosystem members' ability to conduct the job. Therefore, if we are dealing with uninsured workers who have been in an accident and we are interviewing for the constraints, we might ask them, "What kinds of issues concern you in this situation?" "What kinds of things keep you awake at night?" "Are there any issues that are getting in the way of your being able to get the care you need?" This may yield several constraints such as, "I'm worried about the bills and how they will get paid," "I'm afraid that I will lose my job," "I'm anxious about how long it will take to recover and get back to work," and "I'm confused by all the paperwork."

The environmental constraints are also garnered through the interviews but may also be identified through the on-site interviews and simple observation. These constraints include physical barriers, technical issues, political, geographic, and other issues that get in the way of the members' being able to execute their job. For the patients, the physical constraint may be a lack of transportation home after the ER visit or lack of e-mail or cell phone so that the member can be reached for follow-up. For the physicians in this scenario, we might ask them, "What is there about the environment in the ER that keeps you from being able to maximize your effectiveness?" "What kinds of political issues are preventing you from executing your job most effectively?" In response to these, we might hear about such constraints as, "The overcrowding in the ER just generates so much chaos and noise that I can hardly hear my patients, and I'm always rushed; it makes it easy to make a mistake," or, "The politics within the hospital dictate that I admit as many patients that can pay as I can and yet those who can't pay we are trying to prevent from admitting. Of course this isn't discussed, but that's what is happening." These statements yield such constraints as, "Noise and chaos in the overcrowded ER leads to mistakes," "Inconsistent policies regarding who should be admitted prevent me from doing my job right," and so on.

Capturing Outcome Statements for Jobs That Require Extra Detail

When you are gathering outcome statements (which are optional), the interview should be structured according to the flow of the job map. Each of the steps is systematically explored to identify what issues exist concerning speed, stability, and output or yield. When using outcomes as your source of needs, the job step is kept at a summary or high level as opposed to the detailed job steps we create when the job steps serve as the needs. So, if we were producing a job map in the education space for the job of "learn a new subject," the needs generated by the students by the two different methods would look like the data in Table 2-4. (*Note:* in the second column, the

Table 2-4 Outcome Statements versus Job Step Needs

Job Step Category	Job Step (With Outcomes As Needs)	Job Step Needs (No Outcomes)
Define	**Plan to learn a subject** Outcomes (needs): Minimize the time it takes to understand my learning goals, e.g., topics to be learned, skills to be mastered, etc. Minimize the time it takes to determine why I need to learn the subject, e.g., why it's important, how it will be used, etc. Increase the likelihood that I know how my performance will be measured, e.g., what I will be graded on, what scales will be used, etc.	1. Determine the learning goals for a given subject, e.g., why I need to learn the subject, how it will benefit me later, what skills or knowledge I will have to master, etc. 2. Determine how my performance will be measured, e.g., how I will be graded, how much effort is required to get a desired grade, etc. 3. Develop a plan to learn the content, e.g., what to study, when assignments are due, what must be done over time, etc.
Locate	**Locate information sources to study** Minimize the time it takes to determine what will be required to study, e.g., what materials to use, what parts of the materials, etc. Minimize the time it takes to find the sources to study.	4. Locate the material to study, e.g., course materials, additional content on the topic, visuals and diagrams that help explain the topic, etc. 5. Ensure that I have access to the information that needs to be studied, e.g., access to books, Internet, articles, reference materials, etc.

(continued)

Table 2-4   Outcome Statements versus Job Step Needs (*Continued*)

Job Step Category	Job Step (With Outcomes As Needs)	Job Step Needs (No Outcomes)
Prepare	**Determine what will be tested** Increase the likelihood that I know when a subject will be tested. Minimize the likelihood that topics that will be covered on a test are not known. Minimize the time it takes to determine what topics to study to prepare for a test.	6. Determine how testing will be handled, e.g., when, what topics, how frequently, etc. 7. Determine what information should be studied to prepare for the test, e.g., know what topics to study, know how to prepare for the test, how to apply the content, etc. 8. Determine how the test relates to the overall learning goals, e.g., how students will help regain the knowledge, apply the knowledge, etc. 9. Avoid studying the wrong items, e.g., spend time on items not covered, not enough time on topics that are covered, etc.

outcomes are the need statements, and the job step is merely a header or placeholder to contain the set of outcomes in the step. In the third column, the job steps are the need statements.)

As discussed previously in the section on desired outcomes, we are looking for the metrics the customers use as they try to execute the job step. Generally there are three types of metrics that people use to measure the success of a given step: how fast they can get it done (speed), how well they can get it done (effectiveness), and how consistently it produces the right result (consistency). The worksheet shown in Table 2-5 was designed to focus the discussion in order to capture this data for each of the job steps. So in an interview with students using the job map of "learn a new subject," the moderator would ask the student about the job step and confirm that the job step is an accurate depiction of what takes place in that part of the job map. In this case we would confirm that for the prepare step we want to determine what will be tested. Once that is confirmed, we would proceed through the worksheet asking the student, "What makes

Table 2-5 Outcome Capturing Worksheet

Job step	What makes this time consuming?	What makes this inefficient and cumbersome?	What makes this error-prone or ineffective?	Why?

determining what will be tested time consuming?" "What makes this difficult?" "What makes it fail?," and so on. Proficient outcome gatherers would directly translate the students' answers into an outcome statement; this, however, is quite a skill to master and takes considerable time and practice. Therefore, the worksheet can be used to capture the students' answers to the questions, and the outcome statements can be generated following the interview. It is wise to conduct the translation no more than 24 hours after the interview to ensure that you remember the responses well and the context around the responses. If you are using this method, also let the respondents know that they may be contacted to verify some statements if necessary. With practice, the process of dynamically translating the statements becomes second nature, and eventually you can eliminate the need for the documentation followed by translation.

The model can also be used in an asynchronous fashion when it can be sent to the respondent, filled out, and sent back. It can also be used in a Web-based research tool that allows for open-ended responses; this can be a cost-effective means of capturing outcome data across a broad range of people in a short period of time. When the asynchronous method is used, the job steps are presented to the respondents with the four additional columns. The respondents can then answer the statements directly in their own words. When they're finished the worksheet is sent back, and the translation process begins. While this yields a substantial quantity of data in a short period of time, it is important to note the disadvantages of using this method.

First, the interviewer is not able to interact directly with the respondent and therefore does not have the ability to ask probing follow-up questions. Second,

Table 2-6 Need-Gathering Worksheet from Patient Interview

Job Step	What Makes This Time Consuming?	What Makes This Inefficient and Cumbersome?	What Makes This Error-Prone or Ineffective?	Why? What Makes This Job Step a Problem?
Review the patient's history (e.g., patient's chart, medical history, labs and x-rays, etc.).	"Reviewing the patient's history takes time and I generally don't see it until right before I go in to see the patient—especially when patients are back to back; there's just no time to review the next patient." YIELDS: 1. Minimize the time it takes to review the patient's history 2. Increase the chance of having time to review the patient's history before the patient's appointment	"Treatment plans and notes are different for different doctors; makes it hard to follow what took place. Also, sometimes the records are not where they are supposed to be." YIELDS: 3. Increase the chance of understanding notes from other doctors 4. Increase the chance of being able to obtain the record, e.g., it is in the correct location	"Too many times there is key information missing from the chart, e.g., labs, x-ray, etc." YIELDS: 5. Increase the chance that the chart is complete, e.g., up to date, all test results are filed, etc.	Lack of centralized medical records; patients that go from doctor to doctor don't have a common medical history or record. No standardization. YIELDS CONSTRAINTS: • Patients do not have one medical record for all physician visits • Lack of standardization of medical records among doctors

it is not possible to read the translated outcome statement back to the respondent to validate it. This is generally a key part of outcome gathering; however, you will have to decide whether the benefits outweigh the risks based on your specific situation, the goal of the innovation initiative, and so on.

As demonstrated in the completed outcome gathering chart shown in Table 2-6, the commentary can be documented in each column, and the outcome statements can then be garnered from the content. The more traditional way of capturing the statements, however, is to ask the respondent the questions concerning speed, effectiveness, and consistency; identify the outcome statement; and read it back to the member to confirm its accuracy.[5]

Once all the needs are captured from a given member of the ecosystem, they are assembled in a "needs table" which encapsulates all the types of needs in one place. It provides a good overview of the needs of a member of the ecosystem. (See Table 2-7 which shows an extract of a needs table.)

**Ta b l e   2 - 7   Needs Table for Physician
Providing Care, Including Outcomes**

Focal job: Treat Patients

Job Map Step Name	Job Step	Outcomes
Define	• Evaluate the patient's condition	• Minimize the time it takes to obtain a patient's complete medical record including other physicians' notes, lab reports and x-rays
		• Minimize the likelihood that the patient arrives with numerous symptoms other than what he or she is scheduled for
		• Minimize the time it takes to establish a rapport with the patient during the visit
Locate/ prepare	• Make a preliminary diagnosis	• Minimize the time the provider must spend on activities that could be performed by a less skilled, less costly provider
		• Minimize the time it takes to develop a preliminary diagnosis
		• Minimize the time it takes to collaborate with other providers to determine the best course of action (e.g., referral, tests, etc.)

(continued)

Table 2-7 Needs Table for Physician Providing Care, Including Outcomes (*Continued*)

Focal job: Treat Patients		
Job Map Step Name	**Job Step**	**Outcomes**
Confirm	• Obtain diagnostic tests for the patient	• Minimize the likelihood of ordering duplicate services (services that were performed in the past)
		• Minimize the likelihood that a test is ordered because of a lack of time to work through the diagnosis with the patient
		• Minimize the likelihood of ordering a test that the patient cannot afford
		• Minimize the likelihood that the desired tests are not approved by the patient's insurance
		• Minimize the time it takes to receive test results
Execute	• Execute the treatment plan	• Minimize the likelihood that the patient does not comply with the treatment plan
		• Increase the likelihood that the patient takes the responsibility for his or her own care
		• Increase the likelihood that the patient sees a benefit from the treatment plan
Monitor	• Monitor the treatment plan (i.e., how well the patient is responding)	• Minimize the number of physician visits required to monitor the patient's condition
		• Minimize the frequency of patient complications from the treatment plan
		• Minimize the time it takes to determine if the treatment plan is working
		• Minimize the time it takes to respond to patient inquiries during the treatment plan
Modify/ conclude	• Modify the treatment plan as needed	• Minimize the likelihood that modifications to the treatment plan require another office visit
	• Conclude the treatment (record the problem, the treatment plan, what worked, and what didn't)	• Minimize the time it takes to document the course of treatment for future reference

Related jobs
- Manage chronic conditions
- Determine the current health status of the body
- Prevent life-threatening disease at the earliest possible stage

Emotional and social jobs
- Feel respected as a professional
- Be perceived as competent

Human and environmental constraints
- Must practice defensive medicine because of excessive malpractice claims that often require more tests than necessary

Rating Needs and Constraints

Once all the member inputs have been gathered [focal job steps, outcomes (if needed), related jobs, and constraints], the next step is to prioritize them based on their importance and the degree to which the members are currently satisfied with their ability to get the need met. For purposes of innovation, we want to identify the needs that are the *most important* and *least satisfying* for each of the ecosystem members. By targeting this set of needs, we will be able to identify what must be improved for each member of the ecosystem in order to achieve better satisfaction.

For the constraints, it is important to understand which ones the members believe have the most impact on the scenario (and the execution of the focal job specifically) and that they can control.

The methodology introduced by Anthony Ulwick in *What Customers Want* is used to accomplish the prioritization. The quantitative survey instrument has a five-point scale in which the 5 is "extremely important" (or "extremely satisfied"), 3 is "important" (or "satisfied"), and 1 is "not at all important" (or "not at all satisfied"). The lack of a neutral rating in the scale is intentional, thus forcing the respondent to make a choice in one direction or the other.

The questionnaire is developed following traditional quantitative survey techniques with four sections in the questionnaire: screening, profile

Table 2-8 Sample Design for Teachers in an Education Study

Teacher Sample	K–1	2–4	Total
Special education	100	100	200
Gifted children	100	100	200
Standard student base	300	300	600
Total	500	500	1,000

questions, need statements, and constraints. There is one questionnaire created for each of the ecosystem's members since each group will have its own set of needs and screening and profiling questions.

The screening questions are used to identify members who are qualified to respond to the survey. They are also used to find out the questions needed to determine allocation into the sample design, the allocation of the sample across categories. A sample design is shown in Table 2-8. For example, if the scenario involves teachers who work in inner-city schools and teach grades K–4, questions related to these facts would be added to ensure that the right population is recruited. The scenario may also require a focus on teachers who have a certain level of experience, say more than two years, or may want a distribution of teachers that teach remedial, gifted, and standard children.

The next section covers profiling questions. These questions are not intended to assign people to a sample plan nor do they exclude people from the survey; they are used to gather other types of information about the member. For example, for the teacher sample, we might want to know age, ethnicity, level of education, and what kind of technology they use while teaching. This information will allow us to do comparative analyses of needs and constraints based on these characteristics. For example, comparison can be made between the needs of male and female teachers or of teachers who have been teaching for more or less than 10 years.

While the two sections just discussed are found in most questionnaires, the remaining sections are unique to this methodology—the need statements and constraints.

Rating Needs

Needs statements are arranged by job step or are grouped into job categories and are rated for both importance and satisfaction by the respondent. Depending on the size and scope of the initiative, the innovation objectives, and the degree to which the market is understood, there are two ways to approach the rating of the needs.[6] If the program you are executing is using outcomes as needs, then the questionnaire will be structured slightly differently from the way it would be structured if you are using the job steps as needs.

A typical screen for incorporating outcomes looks like the one in Figure 2-6. In the figure, the phrase "scheduling an appointment" is the job step, and the items underneath are the desired outcomes that the member wants to achieve in executing the job step. Respondents rate each outcome for importance and for satisfaction. These surveys can become very long and as such quality controls are put in place to eliminate respondents who are "straight-lining" the survey (basically marking the same answer all the way down the screen or making random patterns on the screen) or those who have not taken the appropriate amount of time to fill it out (basically they took less time to answer than it would take to read it).[7]

The next example demonstrates the method of using the detailed job steps as needs. Remember that with this method the details of speed, effectiveness, and consistency of the job steps will be uncovered during the contextual interviews that will be conducted on the high-opportunity job steps once they are identified.

There are a few different things about this version of the questionnaire. First, only the detailed job steps are listed for rating and often include examples (the "e.g." portion). Second, the "header" statement is changed to reference the focal job for the ecosystem member. In this example, the header is now, "When obtaining health care services . . ." which is the focal job of the patient. As mentioned earlier, it is critical that the job steps be captured as accurate job statements since they are being used as actual

Figure 2-6 Rating Screen for Outcome Needs

	Importance to you					Your satisfaction				
When scheduling an appointment	How important is it to you that you are able to...					How satisfied are you with your ability to...				
	1	2	3	4	5	1	2	3	4	5
Minimize the time it takes to gather information needed to schedule an appointment (e.g., insurance, list of symptoms).	☐	☐	☐	☒	☐	☐	☐	☐	☐	☐
Minimize the likelihood that the provider does not call back with an appointment date/time.	☐	☐	☐	☐	☒	☐	☒	☐	☐	☐
Minimize the time it takes to give personal information when scheduling an appointment.	☐	☐	☒	☐	☐	☐	☐	☐	☒	☐
Minimize the time it takes to describe the health problem when scheduling an appointment.	☒	☐	☐	☐	☐	☐	☐	☐	☐	☒

needs statements. As such, they must be complete, accurate, and adhere to the same rules as job statements.

An important execution note: Even if the intention is to use the job step needs in the questionnaire, it is good practice to capture the outcomes in the qualitative interviews. At a minimum, speed, stability, and output

Figure 2-7 Constraints Rating Screen

With regard to constraints, you encounter as a physician on an optimal health-care delivery system	Constraint's impact						Your control over the constraints				
	How much **impact** is from . . .						How much **control** do you have in solving the problem?				
	1 No impact	2 Little impact	3 Some impact	4 High impact	5 Extreme impact		1 No control	2 Little control	3 Some control	4 High control	5 Complete control
1. Patients want to blame someone when something goes wrong	☐	☒	☐	☐	☐		☐	☐	☐	☐	☒
2. Patients don't take responsibility for their own health and well-being.	☐	☐	☒	☐	☐		☐	☒	☐	☐	☐

statements should be captured for each job step, as indicated in the Outcome-Capturing Worksheet. Capturing this information in the qualitative phase will help you to formulate the detailed job steps and will save you time in the contextual interviews.

Rating Constraints

As mentioned earlier, constraints are assessed based on the impact they have on the social challenge as well as the members' perceived control over them. They are still rated on a five-point scale, however, the scale is changed to reflect impact and control (see Figure 2-8). This information provides a good perspective on issues that must be overcome during the idea generation phase and especially in the execution of the ideas.

The results of this exercise will let us know which constraints each ecosystem member views as high impact and which they believe they can actually control. In some cases, a member may rate a constraint as having a low impact, and he or she may rate control over it as fairly high. If another member of the ecosystem rates this constraint as having a high impact but doesn't have control over it, then this can be an opportunity

Figure 2-8 Impact and Control Matrix for Constraints

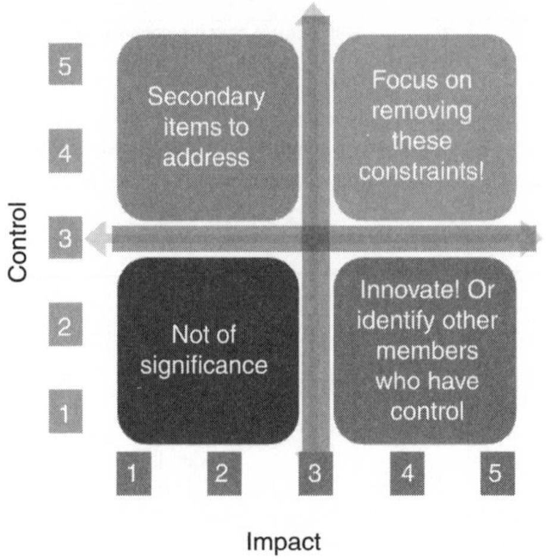

for one member to provide support to another. Figure 2-8 illustrates how we want to approach the constraint data.

For example, in the health-care scenario, among the constraints that physicians consider very important are patients' lack of motivation to do what is necessary to stay healthy. The physician has little control over this. But there is another member of the ecosystem who has much greater control—or at least influence—over patients' motivation: the insurance company or payer. Motivating actions that payers can take (and already do take, in some places and cases) include charging higher premiums when the insured person is overweight, smokes, and the like. For patients who suffer from chronic conditions, payers can also provide information on how to control the conditions (for example, through health fairs, classes, or newsletters) and assign a case manager to the patient.

In this example, it is important to learn the patients' perspective on why they will not take better care of their health. In many cases, obesity, smoking, and other detrimental behaviors have to do with stress. By improving mental health coverage, it might be possible to reduce the problem of noncompliance. In other cases, there is a financial reason for noncompliance: for instance, patients cannot afford the visits to the doctor to monitor their chronic condition. Innovations that target this constraint might, therefore, aim to provide inexpensive ways for people to monitor their chronic conditions.

The Opportunity Spectrum

For each need statement, the primary metric we are looking for is the *degree of opportunity* that exists for creating new value. The High-Opportunity Needs (Hi-Op Needs) will be those in which the member finds the need very important *and* not well satisfied. In the sample completed questionnaire shown in Figure 2-9, these two need statements are "opportunities for innovation" for this respondent since they have high importance and low satisfaction.

This calculation is conducted for each customer need and for each member of the ecosystem in order to arrive at a set of prioritized needs for the ecosystem members. This method is referred to as the "individual opportunity score" because it calculates the opportunity score for each individual respondent and is *represented as a percentage of the sample* that finds the item to be important and unsatisfied.[8] Thus in a sample of 500, if 100 members had the combination of a 4 or 5 for importance *and* a 1 or 2 for satisfaction, then the Hi-Op score for that need statement would be 20 percent. Using the percentage of high opportunity is a straightforward and simple method of communicating the potential for value creation.

Figure 2-9 Completed Questionnaire Indicating "Opportunity"

When obtaining health-care service...	Importance to you						Your satisfaction				
	How important is it to you that you are able to...						How satisfied are you with your ability to...				
	1	2	3	4	5		1	2	3	4	5
Schedule an appointment that is convenient to your schedule	☐	☐	☐	☒	☐		☒	☐	☐	☐	☐
Reduce the time waiting to see the provider	☐	☐	☐	☐	☒		☐	☒	☐	☐	☐

It is also imperative to understand the rest of the spectrum of the need; if 20 percent of the members find the need to be a high opportunity, where do the rest fall? Are they greatly satisfied (i.e., is the market very well served for a vast majority of the members) or are there another 30 percent sitting on the fence with a score of 3 for satisfaction? What is the percent of members who don't find the need to be important at all?

Thus, for each need statement, each member's scores are assessed to determine the combination of scores for the need. Based on the matrix shown in Figure 2-10, scores will fall into one of six categories. For example, a 3 for importance and 2 for satisfaction would yield a low value score; a 5 for importance and 4 for satisfaction indicates well satisfied or high existing value. These categories are further described in Table 2-9.

The individual scores are aggregated to provide a total Opportunity Spectrum for the need statement. The aggregated scores can be calculated for the entire population, or for subsets of the population for comparison purposes such as males versus females or high income versus low income.

Let's look at an example. In Table 2-10 we have five need statements with their percentages as calculated for the total market. These are the scores for patients in a study on health care and the jobs they want to get done in obtaining care.

Figure 2-10 Importance/Satisfaction Rating Results

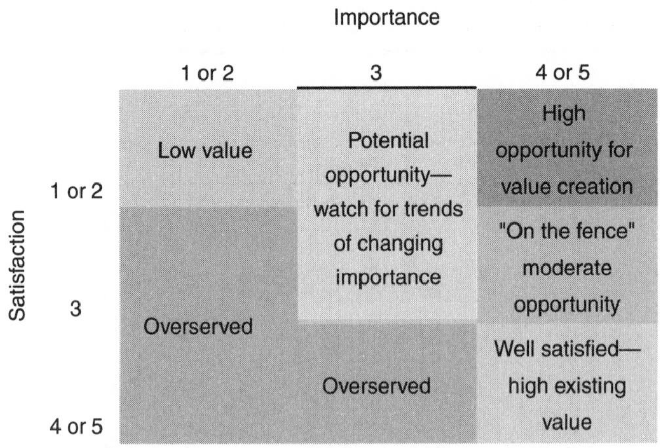

Table 2-9 **Definitions of the Opportunity Spectrum Categories**

High opportunity— focus innovation on these	The focus of innovation efforts should be on high-opportunity areas. These are areas where the members of the ecosystem are in great need of new solutions and improved satisfaction. Those who rate the satisfaction as a 3 we consider "on the fence" as they would likely benefit from new solutions as well.
On the fence— Moderate Op and potential for value migration	Members who find the need to have a high level of importance and only a marginal level of satisfaction are not to be dismissed. Members of this group are on the fence in their satisfaction measures. They rated the item a 3 for satisfaction. In satisfaction score literature, it is frequently stated that those who indicate that they are merely "satisfied" with a solution are those who are most likely to change. The percent of members in this group should be considered as a second level of adopters of a solution that meets the need.
Well satisfied; high existing value	Needs that are already well satisfied and are very important are those that indicate areas of strength in the ecosystem today. It is imperative that the satisfaction of these items be preserved; in other words, we don't want to create solutions that would remove any satisfaction from these areas.
Potential opportunity— watch for trends of changing importance	The area with marginal opportunities should be considered only after the high-opportunity areas are addressed. These items are still important but less so than the others. These may be important to help in understanding when all the needs within the ecosystem are viewed. Because they are rated as less important, if trade-offs among the ecosystem members' needs must be made, these are some areas that could be considered.
Overserved— watch for trends of changing importance	Overserved members rate the need more important than it is satisfied. Some refer to this as "overshooting the mark." The solutions the members have access to today are already not only good enough, they are better than they need to be. These are important items to understand as they can be used to trade off value for improvements in other areas. If giving up some satisfaction in these areas was needed to improve the satisfaction in another area, or for another member of the ecosystem, this is the group of needs that we would target.
Low value	These are needs that for the time being can be ignored. Since they are not important, there is no purpose in improving satisfaction in these areas. Interestingly, in practice it is often the case that some of an organization's initiatives are targeted at improving these needs. This alone is a good reason to execute the program—to understand what not to work on.

Table 2-10 Results Table for Health Care Data

Need Statement—Patients	High Opp	Moderate Opp	Existing Value	Marginal Opp	Overserved	Low Value
Determine what illnesses/ diseases are possible based on the symptoms, e.g., whether a doctor visit is needed, will it get worse without treatment, etc.	9%	39%	30%	3%	17%	2%
Determine how long it will take for the condition to be resolved, e.g., with treatment, without treatment, etc.	28%	38%	20%	4%	8%	2%
Try to ease symptoms at home, e.g., with over-the-counter medications, other therapies, etc.	26%	36%	13%	10%	5%	10%
Identify the best health–care provider to see, e.g., right specialty, experience with your condition, good fit for you, respected physician, easy to talk to, etc.	15%	20%	50%	3%	10%	2%
Find an appointment that meets your objectives, e.g., soon, at the desired time of day, will not miss much work/school, etc.	12%	39%	24%	15%	3%	7%

The combination percentages are calculated for each need statement as just described and reflect the percent of the population that rated the need as the combination in the column head. Thus, in this example, the highest-opportunity needs are, "Determine how long it will take for the condition to be resolved," and, "Try to ease symptoms at home" with 28 and 26 percent, respectively. When the moderate opportunity is considered as well, both of these have over 50 percent of the population in agreement that this need should be addressed. It also appears from this example that the population is already very well served in its ability to, "Identify the best health-care provider," with 50 percent in the well-satisfied category and another 10 percent overserved.

The scores from the worksheet are then illustrated in the Opportunity Spectrum (Figure 2-11) to provide a visualization of the opportunity values.

Figure 2-11   Opportunity Spectrum of Data in Table 2-10

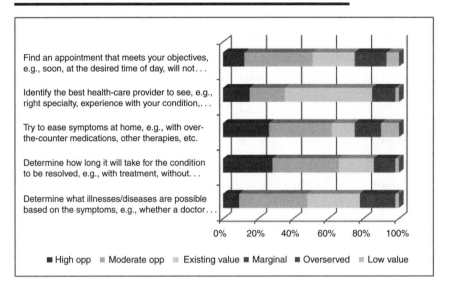

This is especially useful when dealing with a large number of need statements and analyzing the need statements across the ecosystem members.

This technique provides significant insight from the ecosystem members, going beyond the high-opportunity potential and identifying other strategic insights. Knowing which needs are well met today is important when balancing the needs of the ecosystem members in order to: (a) ensure that value is not taken away with new solutions, and (b) knowing what over-served areas are available for potential trade-offs. For those in the government sector who can use this information to make budgeting decisions or a nonprofit trying to determine which programs to put their valuable resources into, this is extremely valuable information. Understanding and studying these different levels of satisfaction is necessary when trying to balance the needs of the ecosystem members.

Segmentation

In many cases, after gaining a sense of the opportunities in the overall market, you will want to dig deeper to see if there are any subsets whose needs align more completely with your capabilities to provide a solution or who provide a more substantial opportunity for creating value. You will already have engaged in a degree of segmentation when you chose what types of ecosystem members you were going to innovate for (e.g., by deciding that you would look at grades K–3 in the education ecosystem), but after seeing the opportunity landscape for your ecosystem members, you may wish to target your innovation efforts more narrowly.

It is likely that you will want to run an analysis to identify any differences based on the profiling questions in the quantitative survey such as demographic, behavioral, or attitudinal. For instance, for the physicians in our health-care ecosystem, we wanted to know whether they were primary care or specialty and whether they were employed by a hospital system or in private practice. We also looked at age, use of personal productivity technology, number of years practicing, and several other details. For the patients, we asked ethnicity, income level, presence of chronic condition, and technology they had at home, among other questions. These questions allowed us to determine if there were any major differences in the market based on simple demographics. We found very different areas of opportunity for primary care and specialty care physicians. Primary care physicians are more concerned with making sure that there is consistency in treatment of the patient by various medical professionals, that there is less waste in the in-patient setting, and that there are fewer delays resulting from authorizations. The specialists are more concerned with getting the patient's complete medical records, with ensuring that the patient sees the benefit of the treatment, and that the patients are held accountable for their own behavior in managing their health and recovery.

In addition to demographic segmentation, a more powerful segmentation can be conducted to identify how the members view the needs differently. This is called Opportunity-Based Segmentation and truly is a segmentation that is performed on differences in opportunity. This segmentation identifies areas of the population that see the need differently and especially if there are segments that have especially high opportunity along a certain dimension or set of needs.

Figure 2-12   Distribution of Opportunity for Homeowners

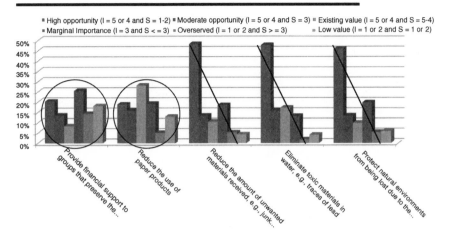

Let's look at some data on homeowners' jobs on "going green," or to put this in a more job-appropriate language, the focal job studied was, "Become a better steward of the earth's resources." (See Figure 2-12.)

In some of the need statements, as illustrated by the circled items in the figure, there is nearly an even spread between those who find the need to be a high opportunity and those who find it either low value or overserved. This shows disagreement among the members as to the significance of this need; it is this type of distribution that drives the segmentation. Items with a linear slope show high opportunity that outweighs all the other categories. These are areas of high agreement and very high opportunity. The numbers that created the figure are shown in Table 2-11.

The output of the segmentation yields groups of ecosystem members that have unique needs. This type of segmentation is very important in studies where there is little high opportunity for the total market and little agreement on the needs. In most social innovation scenarios, we do not find this to be the case; in fact, most of them have so much high opportunity for the total member group that it would take decades to develop solutions to satisfy all of them. For example, in every one of the four health-care studies I conducted while at Strategyn, 80 percent or more of the need statements were high opportunities. This shows significant total

Table 2-11 Data Table for Homeowners

Attribute	High Opportunity (I = 5 or 4 and S = 1 or 2)	Borderline Opportunity (I = 5 or 4 and S = 3)	Existing value (I = 5 or 4 and S = 5 or 4)	Marginal Importance (I = 3 and S <= 3)	Low Value (I = 1 or 2 and S = 1 or 2)	Overserved (I = 1 or 2 and S >= 3)
Provide financial support to groups that preserve the environment	20%	13%	8%	26%	18%	14%
Reduce the use of paper products	19%	16%	28%	19%	13%	5%
Reduce the amount of unwanted materials received, e.g., junk mail, packaging, etc.	48%	13%	10%	19%	4%	5%
Eliminate toxic materials on water, e.g., traces of lead, chlorine, chemical solvents, etc.	48%	16%	17%	13%	4%	2%
Protect natural environments from being lost due to the actions taken by humans	46%	13%	10%	20%	6%	5%

market need. We saw a similar pattern in studies for education. Only in resource conservation did we see big splits in the needs. So the segmentation is valuable and should be conducted if there is little total market opportunity or if the innovation team wants to identify smaller segments in which to target customized programs.

Creating Themes and Finding Synergies

In order to make the ideation process as streamlined as possible, a key step is to group the need statements according to similar "themes." These themes are affinity groupings of jobs or outcomes along some dimension, whether it's their relationship to the focal point of the statement or the object of control. For example, in the theme in Table 2-12 "know the cost" (for a health-care scenario), all the outcomes pertain to being able to know the cost of a service before it is rendered. Theming makes the statements easier to work with in order to identify synergy among ecosystem members. Table 2-12 shows two themes—"manage chronic conditions" and "know

Table 2-12 Themes

Theme	Manage Chronic Conditions	Know the Cost
Outcome statements for workers	Minimize the likelihood that a chronic illness causes more damage to the body.	Minimize the likelihood of not knowing the cost of a test before it is done.
	Minimize the likelihood that the chronic condition gets worse.	Minimize the likelihood of not knowing the cost of medicine before it is prescribed.
	Minimize the time it takes to figure out if a health problem is likely to get worse if left untreated.	Minimize the time it takes to figure out if services can be afforded.
Outcome statements for employers	Minimize the amount of time employees must take off from work for routine health-care visits.	Minimize the time it takes to figure out in advance what the cost of the services will be.
	Minimize the frequency of absences resulting from issues pertaining to the effects of a chronic condition.	Increase the likelihood of knowing the potential cost increase of the benefit program before making the purchase.

the cost"—and the outcome statements that made up the themes. These themes were used across two members of the ecosystem—the uninsured workers and the employers.

It is often the case that the items are grouped based on their affinity to a category or dimension that already exists within your company's vernacular. The example of "manage chronic conditions," is a good illustration of this type of theme as "chronic conditions" are a category that is frequently referenced in the health-care space. Check the opportunity scores of all the items in your theme to get a sense of the strength of the theme itself. For instance, if 25 percent or more of the items in the theme represent strong opportunities, then it is clear that the theme has significant strength and should be pursued. If, on the other hand, only one or two items in the theme offer strong opportunities, then the theme may be prioritized lower when conducting the targeting exercise.

Dynamics of Changing Needs

As time passes and new innovations are introduced, satisfaction of the various needs will also change. This is a given and is why we need to requantify from time to time—to assess how the satisfaction has changed. While changes in satisfaction are the most common, we cannot overlook the fact that importance can also shift, especially when dealing with social issues. Given natural disasters, such as Hurricane Katrina, the devastating tsunami in Indonesia, the earthquake and tsunami in Japan, or problems such as wars springing up in country after country, the importance scores of the jobs of the people will shift accordingly. Before Hurricane Katrina, people's needs were much different from what they were afterwards. Things happen in our environment that can cause significant changes in importance. After the Columbine shootings, security on campus became extremely important, whereas before it was important but not extremely important. When the swine flu epidemic broke out, importance of "protecting oneself from germs while traveling" became extremely important.

Social scenario importance can change drastically because of environmental problems, acts of violence, terrorism, natural disasters, and the like.

Therefore, as we view the opportunity spectrums, keep in mind where changes in importance are likely to occur and always watch the trends of *both* importance and satisfaction over time. This can be done on a scheduled basis (recommended) or as you see significant changes that will affect the scenario. By evaluating changes on a scheduled basis, you will be able to identify trends more quickly and adjust your thinking accordingly; large changes will likely not blindside you as much with this approach.

Once opportunities are identified, we then begin the journey of digging deeper into understanding the *causes* that are driving the dissatisfaction and the importance. Part of this understanding comes from investigating the underlying platform by which the scenario is currently executed, uncovering the problems with the current platform. This information is vital for completing the picture so that new ideas can be generated.

In Summary

As cited in the article on "wicked problems," problem definition is half the battle and is often the reason that the wicked problems go unsolved. Thus the time spent on dissecting the ecosystem, mapping out the innovation blueprint, and exploring different contexts is critical. Once the framing is solidified, then the problem can be articulated and has form and boundaries. It is at this point that the focal job of the members can be identified, the job step clarified, and the constraints understood. To prioritize all these needs, we must get a statistically valid sample so that we can then identify the opportunity spectrum. We will be able to determine which needs are highly underserved, which are overserved, and everything in between. This precision is necessary when attempting to untangle these intractable issues and make real progress in developing solutions that will take hold and make a difference.

Chapter | 3

EXAMINE THE
OPPORTUNITIES

In this, the final "investigation" chapter, the opportunities identified in Chapter 2 are examined to uncover more information about opportunities and to establish the plan for generating ideas about these opportunities. Among all the opportunities identified, those that represent high opportunity for the ecosystem members (Hi-Op needs), are synergistic with one or more other member (S-Needs), those that are in conflict (C-Needs), and the constraints that have the most impact on each ecosystem member as well as across the ecosystem (High-Impact Constraints) are the ones to explore further.

Four Dimensions of New Value Creation

There are four dimensions along which innovation or new value creation takes place—the job, the platform, the job executor, and the business model. It is important to explore each of these dimensions during the ideation process. Table 3-1 illustrates these four dimensions and how they

Table 3-1 Dimensions of Value Creation

New jobs or related jobs	Add other nutrition or herbal supplements to help the child in other ways.	Add new things to Plumpy'nut such as herbal sleep aids at the nighttime meal.
Focal job	Improve the job of providing nutrition via milk products and IV.	Plumpy'nut changes the platform, but serves the same job.
	Existing platform	**New platform**
	Business Model	

would have been used in creating new value in the children's malnutrition area.

A change in job executor often requires a new platform, one that is more geared to the new executor and is usually less complex. Plumpy'nut changed the job executor from the clinic worker (who would administer the IV in the clinic to treat the child) to the parent, and with older children, to the children themselves. This follows the natural course of innovation, bringing the solution closer and closer to the actual beneficiary of the job, eliminating intermediaries, and improving direct satisfaction for the end beneficiary. Business model innovation is discussed in depth in Chapter 5.

Jobs: Synergistic, Unique, and Conflicting

The heart of social innovation is in identifying areas of synergistic opportunity within the ecosystem for which new solutions will significantly improve the members' satisfaction and in turn improve the performance of the entire ecosystem. In virtually every ecosystem, there will be areas of synergistic opportunity. Why is this so? The members are inherently trying to work toward a common job—one that they are all connected to via the social scenario. Teachers and students are both trying to ensure that the student is prepared to be a productive citizen; patients and doctors want to alleviate the medical problem being faced by the patient; first responders and victims both want to prevent death from natural disasters and help the community recover from a natural disaster; governments and citizens

Table 3-2 Theme: Clear Communication

Physician Perspective	Patient Perspective
Ensure accurate patient handoff among treating physicians, e.g., communicate what has taken place, tests given, current medications, etc.	Ensure that all questions are answered during the appointment
Ensure that all physicians who deal with the patient execute the treatment plan consistently	Understand the details of the diagnosis that the doctor provides
Ensure that the patient understands the treatment options	Explain the problem to the provider, e.g., symptoms, what has been taken to alleviate symptoms, etc.

both want to create a safe and productive society in which to live and work. When the overall social scenario is considered, it is clear that the satisfaction of each member is tied to the other members and as such, synergy in the jobs of the ecosystem members should be expected.

In a recent study of the experience of "delivery/receipt of health care," we in fact found several themes of significant opportunity that were present for both the patients and physicians; two of these synergistic themes are shown in Tables 3-2 and 3-3.

As illustrated by these two examples, themes are vital for identifying the areas of synergy and potential conflict. The reason is simple. Different members of the ecosystem are each evaluating a different job within the same "experience." For instance, in the "avoid wasting time" theme, it is clear that none of the need statements are exactly the same, since the two

Table 3-3 Theme: Avoid Wasting Time

Physician Perspective	Patient Perspective
Ensure that there is enough time to spend on diagnostic/treatment issues in order to get an accurate diagnosis	Reduce the number of trips to the doctor, e.g., for follow-up appointments, labs, x-rays, etc.
Spend time in the appointment on patient issues rather than on paperwork	Avoid long waits that result in having to reschedule
Avoid having to repeat tests/procedures	Reduce the overall time spent waiting to see the physician

Table 3-4 Unique Needs

Physician Hi-Op Needs	Patient Hi-Op Needs
• Establish a plan of care for the patient, e.g., personalize the plan based on chronic conditions, comorbidities, etc.	• Try to ease symptoms at home, e.g., with over-the-counter medications, other therapies, etc.
• Determine a preliminary diagnosis, e.g., examine the patient, identify possible psychological issues, etc.	• Determine how long it will take for the condition to be resolved, e.g., with treatment, without treatment, etc.
• Obtain authorization *for the treatment plan*, e.g., provide justification for procedures, medications, equipment, etc.	• Obtain supplies needed for the treatment, e.g., medications, over-the-counter medications, wound care, crutches, etc.

members are doing different jobs. However, by extracting the essence of the job step needs into unifying themes, the needs can be examined across the entire experience.

For each member of the ecosystem, there will also be a set of needs that is not synergistic but that may contain a high-opportunity need for a given member or a high-impact constraint for another member. These "unique needs" must be identified as well. It is important to consider what other needs the members are trying to satisfy, especially when developing the solution.

In the health-care example, there were many Hi-Op needs that pertained only to the physicians, and there were other Hi-Op needs that were unique to the patients. Some of these are given in Table 3-4.

Conflicting needs exist when improvements in the satisfaction of one member's opportunity will have a negative effect on the needs of one or more members. For instance, obviously patients want to reduce their overall costs of health care, especially their out-of-pocket costs. In order to meet this need, either the physician, the hospital, or the payer will have to accept a reduction in fees. This goes directly against the needs of the physicians who feel that they are often not reimbursed enough for the services they provide given their training, internship, residency, and the cost of their education. Therefore, to satisfy the patients' need for lower out-of-pocket costs, the physicians are likely to be unsatisfied with the solution. These types of conflicting needs are often at the heart of many social issues.

Another example in the health-care space illustrates how these conflicting needs can play out. Recent proposed legislation included a mandate that all employers offer health insurance to all employees. While this legislation would satisfy the unmet needs of the uninsured worker and the federal government, it would have significant negative effects on the employers whose needs include "maintain control over how expenses of the company are managed," "ensure that expenses can be covered with the current revenue/expense infrastructure," and so on. Studies showed that the impact of such a move, while reducing the number of uninsured workers, actually would put people out of work! It was projected that there would be significant loss of jobs, as well as employers shifting workers from full- to part time to avoid paying benefits, thus creating a higher negative effect on the employee for whom the legislation was intended to support.[1]

This example demonstrates that if one member of the ecosystem, especially one who plays an overseer role, attempts to make radical changes that will negatively affect the job executors and/or the beneficiaries, there will be significant dissatisfaction across the ecosystem, resulting in conflict and backlash among the members. This type of needs conflict is especially common with overseers and third parties, since their role is often to manage cost, manage the process, and the like, and they have the ability to make decisions that will directly affect the job executor or beneficiaries. It is thus imperative that the *job overseer* understand the needs of the job executors and job beneficiaries before implementing any new programs or solutions. This will ensure a balance of satisfaction across the ecosystem and avoid outright rejection of the solution.

Constraints to Be Considered

The uncovered constraints provide critical information for ensuring that the solution developed will be successful. Without knowledge of these constraints and their impact on the situation, innovators might develop solutions, only to discover an insurmountable barrier upon launch.

Plumpy'nut's amazing success came about by overcoming *all* the constraints that existed in the environment, including a lack of clean water, very little storage space, and the difficulty with carrying supplies over long distances.

In some scenarios, other members of the ecosystem can influence the constraints. For example, one constraint identified in the health-care study was that physicians were working so many hours that fatigue was contributing to the frequency of mistakes being made. Physicians do not have control over their schedules; however, other key members of the ecosystem, the hospital administrator and government regulators, do have control over this constraint. Thus it is important that constraint information be made available to other members of the ecosystem to further their understanding and to help them identify resources they can bring to the table. This is where negotiations begin—and why the unique Hi-Op needs are important. If physicians need a shorter schedule, what needs of the hospital administrators could the physicians help to satisfy in exchange for getting reduced hours?

There are some cases in which high-impact constraints cannot be resolved by any member of the ecosystem. Who then is responsible for overcoming the constraints? You—the innovator. The *solution* must now work to overcome the constraints. In the case of Plumpy'nut, one of the big constraints was the absence of clean water in the areas where malnutrition rates were highest. None of the members of the ecosystem could easily resolve the clean water problem, and yet this constraint directly affected the solutions (milk powder) in the area at this time. In such situations, the *solution* must overcome the constraint by eliminating it, working around it, making the constraint less impactful, and so on. In this example, the solution, Plumpy'nut, eliminated the need for water. Such *uncontrollable constraints* must be resolved within the solution and become a focal point for the idea generation strategy. The topic of overcoming uncontrollable constraints is discussed further in the ideation chapter.

Context

Once the Hi-Op Needs and Hi-Impact Constraints have been identified, it is time to understand the why—both the external factors and internal issues that are contributing to the dissatisfaction. The context data contain vital inputs into the ideation session and in developing business models. As such, it is imperative that the ecosystem members be interviewed directly; this is *not* an area that guessing by the innovation team is good practice unless members of the team have recently worked in the role. Although the innovation team may have some good assumptions as to why the Hi-Op Needs are important and unsatisfied, without asking the actual member, the risk is high that vital bits of information will be missed, and this can have a significant impact on the solution.

When exploring the context, the issues can be studied at the individual need level or at the theme level. The questions in Table 3-5 provide a foundation for understanding the Hi-Op Needs or Constraints that are being investigated. External factors are those to the job beneficiary (such as a patient, a student, etc.) and other environmental factors, and internal factors pertain more to the industry, the policies, processes, and organizational problems that make the problem worse.

Table 3-5 Context Gathering Worksheet

External	Internal
From ecosystem members' perspective	**Organizational industry/sector, or technical issues**
What issues contribute to the problem? What circumstances make the problem worse?	Why has this problem not been addressed by your organization, the industry, or the sector in the past?
Are there members of the ecosystem or subsegments of the ecosystem that are affected more than others?	What has been tried in the past? What worked well? What didn't work?
What work-arounds are used to get the job accomplished?	What processes, technology, people, regulations, etc. make the situation worse?

In the study of health-care delivery we conducted, we found a key opportunity for physicians is to "determine which medication to prescribe for the patient's condition." During the contextual interviews, we identified that the primary issue for the *physician* was the formularies by the health plans. Some of the physicians' issues included different formularies for each health-care system, no centralized area to find what prescription is covered on what formulary, no searchable database (had to look it up on a PDF file alphabetically), and so on. This becomes worse when the patient changes payers since at that point, the process must start over for the *physician* to find out if the drug to be prescribed is on the new formulary.

When we view the problem from the *internal* perspective, from the health-care systems and payers, we find additional information. Among the issues are a lack of Information Technology (IT) budget to make the formulary more accessible and easy to search, and the lack of any industry-wide collaboration to produce a searchable format by health plan. Further probing will be necessary to find out why there has been reluctance to do this and specifically how to move the importance needle for these entities to *want* to make the formulary easier for the physicians to use.

By understanding these issues, the process of generating solutions is substantially enhanced. Why? When generating new ideas to meet a specified need, the ideation team can approach it from one of two angles—direct and indirect.

With the direct approach, innovation teams focus *directly on solving the need* statement, and generate ideas related to that statement. In the example we've been using, the team would begin to generate ideas around how it can improve the way physicians determine which medications to prescribe. One idea may arise where the information is looked up before the appointment by an assistant, or the development of a standardized automated system that can be used to find the information, or some other similar solution. Generally, when approached directly, the teams will focus on just that ecosystem member and how to solve the specific need. This is the case that knowing where the synergies exist can help broaden the perspective of the ideation teams and have them look at solutions that are ecosystem-wide.

With the indirect approach, the focus turns to *resolving the underlying issues* identified in the need statement. This often leads to a more holistic approach that engages other members of the ecosystem. In our example, instead of focusing on how a physician can better determine the right medications via resources in his or her own practice, the first step instead may be to determine how to get the health-care system to contribute to a centralized formulary database. This can be accomplished through a government agency or private corporations, that would create the centralized formulary system. In the case of a government solution, the data would then be available to the public; in a private corporation, the system could be resold to physician groups around the country, or world, and could be sold to health plans for their members to be able to easily look up a medication.

Defining and Utilizing "Value Delivery Platforms"

A *value delivery platform* (referred to as "platform" going forward) is the means by which value is created and delivered to the target audience. It includes the capabilities, infrastructure, people, processes, and resources of the ecosystem members. All products, services, and programs have a platform on which the solutions are delivered. The platform is composed of the system infrastructure and subsystems that deliver the core product or service function and which are vital for innovation to take place.

Let's look at each of the components of the platform in more detail.

- *Infrastructure:* For a physical product, the infrastructure is the core foundation by which the value is delivered and includes the materials the product employs, its energy source, and its size and shape. In services and programs, the basic components of the system are less prevalent and are more likely to be focused on processes, people, and technology.

- *Subsystems:* Core mechanical, electrical, chemical, and/or software components for physical products. For social situations this might be localized systems, business models, service mechanisms, and the like.

- *People:* The people involved in selling, servicing, creating, and executing programs, services, or products.

- *Technology:* Any special technology used to create or deliver the product, especially that which is unique to the industry, a company, or a given delivery mechanism.

- *Process:* In social scenarios, process is a big part of the platform and refers to any of the special processes that make the system work today in delivering the service to the customer. This can include fundraising, support services, program delivery, and so on.

When we consider the phrase "thinking inside the box," it is generally the case that people are stuck within the confines of the existing platform. The platform is a very substantive "box" that binds our thinking to the way things are done today. When people refer to "breakthrough solutions," they are often thinking of breakthroughs that take place at the platform level. Netflix, iPhone, Facebook, and dozens of other major innovations were revolutions in the *platform*. In the nonprofit world there are also dozens, although perhaps less well-known, including the Nature Conservancy's Water Trust Fund (the new program that brought fresh produce to food banks in southern California) and even Plumpy'nut. These are all examples of using new platforms to accomplish the goal of the organization. It is for this reason that we must identify the facets of the current platform so that we can begin to understand whether they are the best way to address the needs or whether they are actually part of the problem in people getting their needs met. In the health-care system, one of the major subsystems in the platform is the insurance company which is often seen as part of the problem in obtaining health-care services and inhibits the successful execution of new platforms.

The combination of these elements provides a platform on which jobs are executed. Let's take several examples starting with a simple

Table 3-6 Platforms in "Removing Food Particles from Teeth"

Job: Remove food particles from teeth

Platform	Core Aspects of the Platform
Standard toothbrush	Contains fixed bristles on a plastic handle. Removes particles through motion of rubbing the bristles against the teeth and gums; the motion is generated by human effort.
Electric toothbrush	Contains a motor that holds rotating bristles (change in subsystem by adding the subsystem of the motorized unit). Removes particles mechanically using energy from the motor as opposed to human effort.
WaterPic	Contains a pressurized stream of water (complete change in infrastructure from mechanical) that is directed toward food particles, dislodging them from their location.
Future: Chewing gum	This gum contains a chemical that dissolves food particles found in the mouth (complete change in infrastructure, subsystems, and process).

physical product to get a feel for the concept. For the job of removing food particles from the teeth, the standard platform for generations has been the basic toothbrush composed of a plastic handle, bristles, and so on. While some companies are still improving the basic platform of the toothbrush by adding more functionality such as tongue cleaners and cheek cleaners, new platforms have arisen to challenge that platform including electronic toothbrushes, tools like the WaterPic, and even gum that dissolves food particles. Table 3-6 shows several different platforms that address the job of removing food particles from the teeth.

To take this even a step further, a more dramatic change in platform is found in the animal kingdom where the Egyptian plover (bird) cleans the teeth of crocodiles.

While we used the toothbrush as a very basic example, it is obvious that social scenario platforms are significantly more complex with many more facets within the platform elements. Let's look at the health-care space as a typical example of a social scenario. As demonstrated in Table 3-7, the health-care platform is composed of dozens of components that combine

Table 3.7 Health Care Delivery Platform Elements

	Components of a Traditional Health-Care Delivery Platform
Infrastructure	Hospital systems, urgent care centers, independent practicing physician offices, pharmacies
Subsystems	Specialty care centers within hospitals, triage units, labs, insurance companies, government payers, hospices, rehab facilities, nursing facilities
Technology	Medical records, imaging and diagnostic technology, advanced technology for caring for major illnesses and diseases, pharmacy IT systems, payer systems, government systems for billing
Processes	Scheduled appointments; primary care recommends to specialists, type of facility to go to is determined by the patient but sometimes dictated by the insurance payer, copayments and coinsurance, insurance purchase and use process, diagnostic coding, credentialing process
People	Physicians, hospital administrators, physician assistants, lab technicians, pharmacists, student doctors (in teaching hospitals), payer employees (medical oversight, credentialing, marketing, case management, etc.)

to create the standard platform. When we consider all the various jobs we expect the health-care system platform to perform—from treating terminal illnesses to fixing broken bones and alleviating colds—it becomes readily apparent that one platform for such different jobs is highly inefficient. The current health-care delivery platform is built for the more intensive health-care jobs and as such is overkill for the smaller jobs such as a cold, the flu, or a sprain. We discuss why this takes place and what is being done about it in Chapter 7.

When a new platform emerges, it is easy to identify how the platform has been changed to more efficiently address the needs of the customer. Table 3-8 compares the platform components of the standard health-care platform with two other platforms.

Plumpy'nut was also a change in platform. Prior to the introduction of this new platform for treating malnourished children outside a hospital setting, Nutriset and other organizations delivered a product called F100 consisting of a powdered milk formula which required clean drinking water,

Table 3-8   Alternative Health-Care Platforms

Job: Treat a Minor Illness

Platform	Core Aspects of the Platform
Standard health-care platform—doctor visit	Appointment scheduled, visit with doctor—usually MD—obtain prescription, go to pharmacy to get needed supplies, insurance covers visit—patient usually covers copay
Innovative platform: MinuteClinic	Walk-in visits available, visit with nurse practitioner or physician assistant; prescriptions provided and filled at the store in which the clinic is located; usually paid for out of pocket and not through insurance for about the same price as the copay
Innovative platform: WebMD	Self-evaluation of condition, no appointments, immediate access to information, no cost to use

precise measuring, and mixing. It also had a very short life span, requiring consumption very soon after it was mixed or it would lose its nutritional properties. Table 3-9 illustrates the differences between these two platforms.

The benefit of understanding the platforms involved cannot be overstated. As we'll see later in the chapter on idea generation, it is the platform itself that provides the most significant potential for innovation either through modifications of the platform or through resources found on the platform that can be used to create new features.

Table 3-9   Platform Innovations in Plumpy'nut

Job: Treat a Malnourished Child

Original Platform: F100 Milk Powder	New Platform: Plumpy'nut
• Powder formula to mix with water	• Peanut-based substance with no water or mixing required
• Requires parent to mix and prepare in a vessel to feed the child	• Vacuum packed in a way that enables the child to squeeze the package to get the food out
• Must be used immediately	• Shelf-stable for two years or more

Developing the Ideation Portfolio

With all the information now in hand, it is time to structure the *ideation portfolio*, the overall plan for how the opportunities will be addressed. It is composed of a *series of "Ideation Strategies"* that are designed to systematically address the opportunities, along various platforms and time horizons and to address various stakeholder criteria. Each specific Ideation Strategy is defined by several factors:

1. *Members:* Which member or members of the ecosystem will be focused upon? Are there specific subsegments of the members that will be targeted?
2. *Opportunities:* Which opportunities will be targeted? In many cases there will be more opportunities than can be handled at one time.
3. *Platform:* Can the opportunities be addressed on an existing platform? If so, which one, or does a new platform have to be created?
4. *Time horizon of the solution:* Do we need quick wins or medium-term solutions? Or are we looking for breakthroughs that can take years to implement?
5. *Stakeholder criteria:* What are the priorities of the organizations' key stakeholders? Speed to launch, low cost to the end user, low development cost and effort, and/or maximization of the social value created?

By defining the Ideation Strategy within this framework, it becomes feasible to generate solutions that will address many different yet concurrent strategies. Figure 3-1 illustrates a sample of an idea generation portfolio that will yield several different concepts based on different opportunities, different audiences, different platforms, and different management criteria.

Let's look at an example. Nutriset, armed with information about the needs of aid workers, parents of malnourished children, physicians, and so on could establish several different ideation strategies as shown in Table 3-10.

Figure 3-1 Idea Generation Portfolio

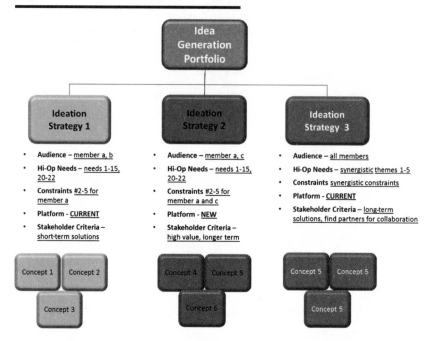

These different Ideation Strategies would have yielded different solutions, primarily because of the focus of the platform. In strategy one, the focus was to improve upon the current platform and the time horizon was shorter term. Given the requirements, it is possible that addressing the constraint of removing need for water can be done through the existing platform and may need a new platform such as a biscuit with the F100 formulation.

While this is a fictitious example (we don't know that Nutriset employed such a process in developing these products), it is easy to see how simple changes in the Ideation Strategy can result in very different products. By focusing ideation teams on specific opportunities within specific time horizons on specific platforms, very different results will be obtained. This is the ideal scenario—pursuing the problem from several angles and generating multiple concepts to choose from moving forward. This approach also provides the organization with a portfolio of concepts that are in different stages of development, thus optimizing the frequency of innovation

Table 3-10 Ideation Portfolio

	Ideation Strategy 1	Ideation Strategy 2
Target customer	*Parents* of malnourished children; *same as current customer base*	*New customers* in rural areas as well as current customers
Prioritized needs and/or constraints	Eliminate need for adding water	Eliminate need for adding water
	Reduce the preparation time	Reduce the preparation time
	Allow for self-feeding	Allow for self-feeding
		Improve the efficacy of the product—enhance the nutrition of the solution
Platform	Current platform and formulation to ensure rapid market entry	New platform desired to move away from the powdered milk
Time horizon	Short term	Long term
Stakeholder criteria	Low risk, low effort to develop	Ability to reach more people in rural areas and increase performance of product (social value)

launches, program enhancements, and program expansion. These create a *concept road map* that will enable the organization to release new concepts on the existing platform, new platforms and even create disruptive solutions over a period of years. (See Figure 3-2)

Figure 3-2 Concept Road Map

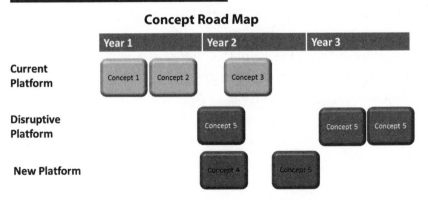

Companies often do not use this approach. Instead they conduct ideation sessions as a one-time event, attempting to cover all possibilities. This results in a hodge-podge of ideas including new long-term platform ideas, quick-win ideas on existing platforms, new business models, and so on. The problem with this approach is that these ideas are then assessed against each other to determine which ideas to pursue. Because the ideas are a mix of many different kinds and because each has a different objective, it is difficult at best to make comparisons among the ideas. In such cases, the concepts selected are usually the short-term, easy-to-implement ideas, resulting in incremental innovation. For breakthroughs to take place, the team must be directed and given permission to make such a longer-term, riskier choice. Using the Ideation Portfolio ensures that both paths are pursued.

Using the portfolio allows organizations to create a concept road map which yields continuous innovations over time, creating some quick wins to energize the team and encourage the stakeholders as well as producing breakthrough ideas that can truly change the social landscape.

Thus the concept road map is the result of the execution of the ideation portfolio in which one strategy approached short-term solutions with the current platform. One group focused on a disruptive platform and another on an entirely new platform.

The Ideation Strategy Guide

Once the ideation portfolio has been developed, there will be multiple Ideation Strategies within it. Each Ideation Strategy must be documented and detailed to ensure that the innovation teams have clear direction on how to proceed, what they are expected to produce, and any other requirements. The Ideation Strategy Guide (see Figure 3-3) is created for each of the strategies within the ideation portfolio identifying which opportunities and audience will be targeted for that particular strategy, which platform(s) will be used or considered, what time horizon

Figure 3-3   Sample Ideation Strategy Guide

Ideation Strategy Guide		Date:
Targeting Objectives		

Solution Timeframe	Platform	Stakeholder Criteria (check top 3)
❏ Short term _____ ❏ Medium _____ ❏ Long term _____	❏ Existing platform ❏ New platform creation ❏ Business model creation Details: _____ _____	❏ Low cost to end user ❏ Low development effort (resources and money) ❏ Speed to implementation ❏ Risk – must have low technical, political, organizational, or societal risk ❏ High ability to create social value ❏ High ability to diffuse the solution to many locations

Member	Target Segment & Size	Opportunity Level	Rationale

Synergistic Needs

Unique Needs or Conflicting Needs	
Member 1	**Member 2**

Constraints to Target		
Human – Member 1	Human – Member 2	Environmental

is desired, and which stakeholder criteria will be used to prioritize the concepts.

There are two key purposes for using an Ideation Strategy Guide. First, the guide helps to maintain the focus of an ideation session. It ensures that the ideation team will be focusing on the right perspective, the right platform, the right audiences, and the right opportunities. Second, having

Figure 3-4   Ideation Strategy Guide for Education

Ideation Strategy Guide–Example

Targeting Objectives

Solution Timeframe	Platform	Stakeholder Criteria (check top 3)
☑ Short term (6–12 Months) ☐ Medium _____ ☐ Long term _____	☒ Existing platform ☐ New platform creation ☐ Business model creation Details: _____	☐ Low cost to end user ☐ Low development effort (resources and money) ☒ Speed to implementation ☐ Risk – must have low technical, political, organizational, or societal risk ☒ High ability to create social value ☒ High ability to diffuse the solution to many locations

Member	Target Segment & Size	Opportunity Level	Rationale
Students – high school	Inner-city, public school attendees	Significant	This group has significantly higher opportunities in a few critical areas
Teachers – high school	Inner-city, public school teachers	Extreme	Same focus to develop synergistic programs and solutions

Synergistic Needs to Target

1. Gain access to the technology needed to succeed in today's jobs
2. Ensure students graduate from school with the knowledge and skills they need to get a good job
3. Ensure a safe environment for the kids to learn
4. Arrange financing for college for low-income students

Unique Needs or Conflicting Needs to Target

Students	Teachers
Prove myself to my peers Avoid being "noticed" by dangerous groups Determine how to get school work done while working to help support the family	Show support for the kids Encourage children that they can accomplish more than what they currently have at home

Constraints to Target

Human – Students	Human – Teachers	Environmental
Home environment of many students is unstable, stressful, and sometimes abusive Hard to study when working to help support the family	Kids don't take school seriously Many kids are exhausted because they work full time to support their families	Inner-city schools are often under-funded, do not have the technology of the schools in "wealthier districts", there is high turnover in teachers, there is a very low graduation rate as kids drop out to work to make a living, too high student-teacher ratio making it difficult to give the kids the time and help they need

2

the Ideation Guides allows multiple sessions to take place simultaneously, each one focused on its own set of objectives and resulting in concepts that are designed to be launched over a period of time.

As mentioned previously, there may be several Ideation Guides generated from a single program. For instance, in the case of the education initiative, there may be one Ideation Guide focusing on in-classroom needs, using opportunities from teachers and students who are in a segment of

Figure 3-5   Social Innovation Ideation Framework

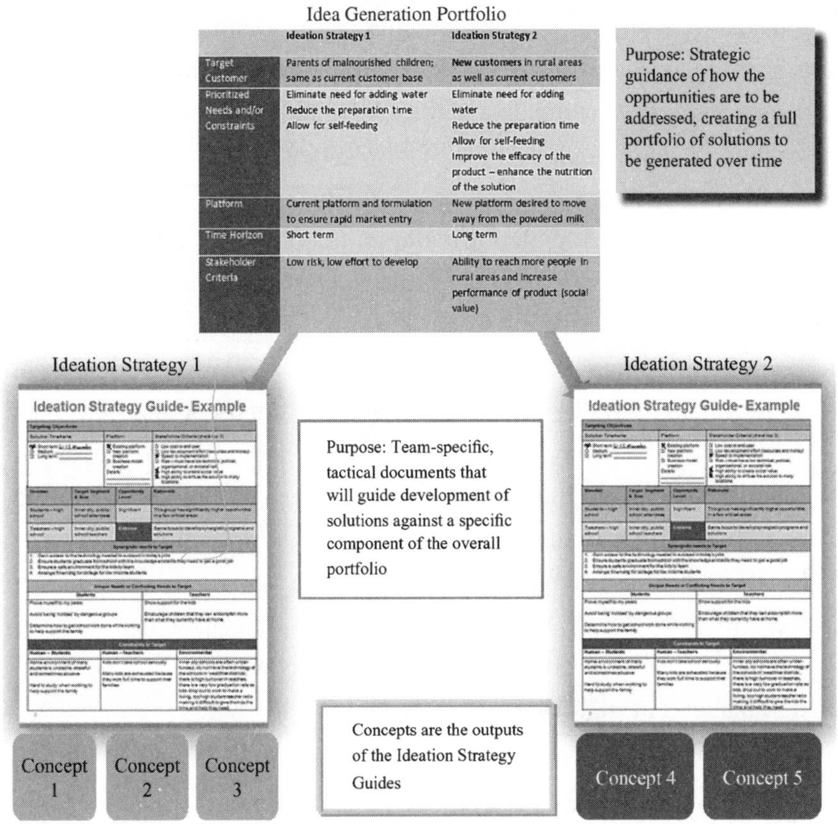

Idea Generation Portfolio

	Ideation Strategy 1	Ideation Strategy 2
Target Customer	Parents of malnourished children; same as current customer base	New customers in rural areas as well as current customers
Prioritized Needs and/or Constraints	Eliminate need for adding water Reduce the preparation time Allow for self-feeding	Eliminate need for adding water Reduce the preparation time Allow for self-feeding Improve the efficacy of the product – enhance the nutrition of the solution
Platform	Current platform and formulation to ensure rapid market entry	New platform desired to move away from the powdered milk
Time Horizon	Short term	Long term
Stakeholder Criteria	Low risk, low effort to develop	Ability to reach more people in rural areas and increase performance of product (social value)

Purpose: Strategic guidance of how the opportunities are to be addressed, creating a full portfolio of solutions to be generated over time

Ideation Strategy 1 — Ideation Strategy Guide- Example

Ideation Strategy 2 — Ideation Strategy Guide- Example

Purpose: Team-specific, tactical documents that will guide development of solutions against a specific component of the overall portfolio

Concept 1   Concept 2   Concept 3

Concepts are the outputs of the Ideation Strategy Guides

Concept 4   Concept 5

unwilling learners, which will use existing platforms (existing classroom structure) and will focus on medium-term solutions. Another Ideation Guide may be focused on opportunities of related jobs outside the classroom, focusing on parents and students, new platform development, and a longer-term timeline. Figure 3-5 illustrates how the Idea Generation Portfolio may require several Ideation Strategy Guides to deliver different types of concepts.

Before continuing on the setup of the ideation, I'd like to pause and provide a demonstration of the various types of solutions that can arise based on the type of platform being used.

Platform Types in Ideation

The platform is one of the key elements that distinguishes the Ideation Strategy Guides as it will have significant impact on the output. There are three options available; the data will usually guide which one to use. The options are (1) existing platform with new features, (2) low-end disruption where the new platform is less complex and less expensive, and (3) a high-end disruption where the platform allows the job to be performed better but it also may be more expensive. If the data show significant overserved needs, then it is likely that a low-end disruption is the appropriate play. If the data show that much of the high opportunities cannot be addressed on the existing platform, then it is likely that a new high-end platform must be designed to accommodate the unmet needs. If most of the needs can be addressed by the existing platform and the goal is a short-term solution, then adding features to the existing platform is the appropriate strategy.

To illustrate the difference among the three approaches, we use an example from the health-care field (see Table 3-11). The audience in this case is patients who are struggling to manage a chronic health condition, such as diabetes or high blood pressure.

Existing Platform Scenario

For the existing platform scenario, the team is asked to provide better access for these patients who must see the doctor on a regular basis using the existing health-care platform consisting of physicians, hospitals, traditional payers, and the overall health-care infrastructure. An idea that meets the need of frequent physician checks of their vitals is group appointments, also known as shared medical appointments. During these appointments, the patient is seen with other patients who have the same condition. This new type of medical appointment is being used with chronically ill seniors, high utilizers of service, those with a specific diagnosis who require frequent monitoring, and so on. As described by the American Academy of Family Physicians, "These visits are voluntary for patients and provide a

Table 3-11   Needs of Physicians and Patients on Chronic Conditions

	Physician	Patient
Needs	Manage chronic condition of patient	Manage chronic condition
	Determine how patient's vitals have changed over time	Reduce the number of visits needed to monitor the condition
	Observe patient's condition at a nonacute time	Determine how the condition has changed over time
Constraints	Do not have enough time in the schedule to see all patients as often as they should be monitored	Not all specialists are available within a reasonable driving distance

secure but interactive setting in which patients have improved access to their physicians, the benefit of counseling with additional members of a health care team (for example, a behaviorist, nutritionist, or health educator), and can share experiences and advice with one another."[2] The physicians treat each individual within the group based on his or her unique situation, while other team members such as nurses, social workers, psychologists, and behaviorists answer questions and lead group discussions on relevant issues. The appointments are 90 minutes long, allowing much more time for the patients to receive information about their condition, get questions answered, and gain support from others with the same condition (which are all Hi-Op needs of the patients). Each patient also receives direct feedback on his or her specific medical condition through the evaluation of blood work and tests that were ordered and completed before the appointment. This type of appointment reduces the physician's overall time, spreads the cost over several patients, and allows the patients to care for their chronic condition at a lower price.

Low-End Disruption Scenario

If the data suggest that the patients with chronic conditions are already overserved by the existing health-care system, then to get the job done of monitoring their chronic condition, the team can focus on a low-end disruption. An example is retail clinics like the MinuteClinic model available through CVS Pharmacies. This is truly a low-end platform with a small medical office located in retail space, staffed with nurse practitioners or physician's assistants, and handling only low-complexity medical conditions. A study conducted by the *Annals of Family Medicine* in 2010 found that while most patients preferred to see physicians at their office practice, they would seek care at a retail clinic based on more immediate appointment availability and price. The study found that appointment wait time was the most vital factor in the decision concerning where to seek care.[3] This explains the attractiveness of this low-end disruptive model where walk-in appointments are welcome and are the norm.

High-End New Platform Scenario

If the data do not show that the members are overserved but find that they are very frustrated with having to go to the doctor just to get their frequent checks, then a high-end disruption solution would be successful. A recent NPR (National Public Radio) special showcased a physician who treated Parkinson's patients through a videoconferencing setup at a local nursing home. Patients would come to the center to access the videoconferencing unit, have their appointment with their physician, and then handle any follow-up such as getting new medications filled, and the like. "Remote access to medical care has been touted as the next great thing for almost 20 years. And telemedicine is now more widely used in some areas, such as linking radiologists and stroke specialists to hospitals. It's been used to monitor patients' vital signs remotely, and to provide long-distance psychiatric care."[4] This high-end new platform is well underway to becoming a breakthrough in medicine; however, like many new platforms, it is not financed as the previous platform is. Insurance programs often do not cover the costs, despite their effectiveness;

thus the patients must pay for this new platform out of their own pocket. For the benefits it provides, many are willing to do so.

These three examples illustrate the need to understand the type of innovation to be generated by the team. Without knowing where to focus, ideas ranging from the retail clinic, to group appointments, to video medicine would all be generated. It would be difficult to choose from among them because they all have value depending on what the objective of the innovation program is and where the needs lie. With the approach of using an ideation portfolio with several different Ideation Strategies, it is possible to generate ideas in each of these venues, giving the organization substantial breadth of possibilities for launching a breakthrough solution in the long term and immediate wins in the short term. Keep in mind that if the goal is to create a new platform, sessions to address the new platform should precede feature-level ideation for obvious reasons. The feature-level ideation team needs to know what platform features are being added onto.

Technology and Competitive Solution Scan

The last bit of information needed before an effective Ideation Strategy can be executed is a scan of existing, new and pending intellectual property—either within the organization, across multiple organizations, or across an entire industry. Even in social scenarios, these technology and competitive scans are important to identify potential off-the-shelf solutions that can be purchased, companies that can become partners in the endeavor to solve the problem, and potential acquisitions in some cases. As we will learn later, one of the core precepts of the TRIZ problem-solving techniques is that someone somewhere has already solved a similar problem. The most prudent step then is to find out where and when that problem was solved and determine how that idea can be applied to the current situation.

In the past couple of decades of conducting ideation work, it never ceases to amaze me how the innovation teams are *not* aware of existing technologies, sometimes even within their own company. I worked with a group that made pharmaceuticals for cancer patients, and the teams were excited, had great information about the needs, the context, everything they needed. They came up with some truly breakthrough ideas, at least what they thought were breakthrough. As they presented the ideas to management a few days later, they were astounded to learn that this same idea was already well into development by their own company in another division halfway across the country. This happens frequently and is a real bummer for the innovation team. Thus, after having this happen one too many times, we began to include a technology scan of current solutions in the market, new solutions just launched, new Intellectual Property (IP), pending patents, and internal patent and idea searches. This became an extremely beneficial exercise since the teams were made aware of what already existed or what was pending, and they could use those items to launch new ideas from, build on the ideas.

The following example demonstrates the effectiveness of technology scans in generating new ideas. The group Doctors Without Borders has to be highly innovative in its work because its members are often working under substandard medical conditions fighting some of the world's most deadly diseases. This group has produced some innovations that help its members in the field such as a simple and rapid test for malaria that does not require a laboratory to obtain results, a fixed-dose combination drug for AIDS and malaria that provides better dosing control and a single tablet for the patients (increasing the chance that they will take the medication), and the use of digital medical imaging in the field.[5] A technology scan that identified these innovations would be very beneficial to a team working on innovations to improve medical care immediately following natural disasters. A review of the innovations generated by Doctors without Borders would be a prudent step, not only for a direct transfer of the innovation, but for the process by which the problem was solved. The process could shed light on similar situations being faced by first responders and medical professionals after natural disasters where entire towns are destroyed or disabled for an extended period of time.

Technology scans can be conducted within the public domain using tools such as Google or patent databases, and they can be conducted within one's own organization to explore potential solutions that the organization already has available and that might address the targeted opportunities. This applies not only to large corporations, but to government agencies, large nonprofits, and other widely distributed organizations which are often unaware of solutions that exist within their own company. This is often the result of a lack of a common, centralized repository for innovations, ideas that were not implemented, field notes of ideas, and so on.

Keep in mind that the technology scans should be conducted based on *opportunities*, not necessarily just on the obvious current products, technologies, or solutions. For instance, if we are trying to solve a problem concerning communicating visual information to others, we might consider Microsoft PowerPoint, or Visio and other software as obvious technologies. But we must also consider that the alternative for *getting the job done* may simply be a pen and paper or a camera in a smart phone. Thus in conducting the scan, we would want to explore other solutions for communicating visual information to others and not focus on existing technology only.

In a recent study on health care in the home, we came across dozens of new technologies that are under development to provide care in the home. However, many of these technologies are in their infancy and are quite cost-prohibitive. Therefore, it is important to not only seek out cutting edge technologies but also to explore those that are at an affordable place in their growth cycle.

Another strategy for teams looking for a disruptive play is to find new uses for old technologies. In such cases where the needs are overserved, then the older solutions, with less complexity, features, and cost, may become a good solution to address these needs. For example, Doctors without Borders often uses old generation medical equipment to conduct jobs in the field where "good enough" is better than "nothing at all."

In Summary

As this chapter illustrates, obtaining the opportunity spectrum scores is just the beginning of the investigation. Once the Hi-Op Needs and Hi-Impact Constraints are identified, there is still much to be done: identifying synergies and conflicts among the ecosystem members, identifying the causes behind the dissatisfaction, determining on which platforms the needs will best be addressed, and creating the ideation portfolio that will guide the ideation process for months or even years. After this, the investigation phase ends, and we are off to idea generation, business model development, and diffusion of innovation strategy development.

Part | 2

INNOVATE THE SOLUTION

Chapter | 4

DEVISING A WORKABLE
SOLUTION

You may be wondering why it took so long to get to generating innovative solutions in a book on innovation. As the title of this chapter indicates, we are looking for a *workable* solution; therefore, it is imperative to obtain all the right inputs at the start—*before* ideas are generated. At Strategyn we often explained to our students and clients that starting with ideas first is like trying to solve a simultaneous equation by *guessing the answer.* Take equations such as $2X + Y = 3$ and $X - Y = 4$ and try to solve it by guessing that $X = 2$ and $Y = 1$, and that $X = 5$ and $Y = 3$. This could go on forever. Such a task would take an extraordinary amount of time and, worse yet, it is very possible that after all that guessing, you may still not ever get it right. That's why there is a formula and a process for solving such equations. Innovation can, and should, be approached in much the same way—as a methodical process. As we discussed earlier, Thomas Edison used a process and approached innovation from a needs-first approach and found it to be highly successful.[1]

Within the simultaneous equation, the "needs" of the ecosystem members are like the constants because they are stable over time. Thus, once these needs have been identified, it is possible to systematically solve for the variables, the solutions, which continue to evolve over time to better meet the customers' needs. Our goal in social innovation is to take the solutions that exist today to the next level, addressing as many needs across the ecosystem as possible and thereby elevating satisfaction for all members. It is at this point that we begin to see real progress in solving social problems.

A word about creativity—what it is and what it is not. I've heard many people say that they are just not creative. I've found that when they say that, they mean that they cannot come up with a breakthrough idea out of thin air that will meet a great market need. Most people can't—with the rare exception of folks like Steve Jobs and Bill Gates—create great ideas on the fly with regularity. For those of us with a more modest allocation of creativity genes, the right information and the right structured process can yield significantly creative insights.

I've seen over the past two decades that providing innovation teams with the customers' needs and a framework that guides them through a structured process for breaking through their traditional modes of thinking ensures that *everyone* can be creative.

Interestingly, for many of the organizations I've worked with, the problem is usually not an absence of ideas, but an absence of knowing which are *good ideas*. This is precisely why we spend so much time on understanding the problem *before* we attempt to solve it and why the ideas-first method of innovation is so dangerous. Organizations that are tackling major social problems and have limited resources to do so cannot afford the "fail faster method" of innovation. This fail faster theory has gained momentum in the corporate world over the last several years and the thinking goes something like this: we generate ideas and send them into the market; we keep a close eye on how well the market likes them; we kill the bad ideas quickly to avoid losing money. This is an attempt to fix the problems inherent with the ideas-first approach. We believe instead, that organizations should simply reverse the process, get the needs first, then generate solutions to meet those needs and then there will not be a reason to fail faster—our goal is to not fail at all. This is especially important in social innovation where organizations *must* be able to hit the target the first time every time. These organizations need to have a consistent and reliable process for generating innovative solutions. Such was the impetus for writing this book.

Let's look at all the information we have available at this stage of the process. With respect to the Hi-Op Needs of all ecosystem members, we know:

1. The Hi-Op needs of all ecosystem members

2. Which Hi-Op needs are synergistic among the members and which are conflicting

3. Which customer segments should be targeted among the ecosystem members

4. The High-Impact Constraints (human and environmental) that must be overcome

5. The root causes and issues surrounding the Hi-Op needs

6. The stakeholder criteria, such as what defines success in the minds of the innovation team and the people they report to

With this wealth of information, success is driven by two factors. First, all of the team's energy is put to work on exactly the right problems—the unmet member needs. Thus the team is highly focused, maximizing its effectiveness at hitting the target. Second, ideas that are generated can be immediately assessed against the needs of the ecosystem members in a virtual lab setting which provides immediate feedback as to the strength of the ideas. No more guessing as to whether an idea is good or bad. No more politics to push an idea through. It is a good idea if it addresses a Hi-Op Need directly (or indirectly) or overcomes a Hi-Impact Constraint, and is within the guidelines of the stakeholders. Other ideas really are bad ideas. It all comes down to the effective use of boundaries.

Ideation Boundaries

Let's face it. In all brainstorming, creativity sessions, ideation sessions, whatever your organization calls them, there are real boundaries. You can choose to suspend the boundaries for a period of time to give people the freedom to create any idea, an approach sometimes called "blue sky," but that often results in frustration of the team members when their blue sky idea is shot down because it didn't meet some invisible criteria of the management team. It is much more effective for the team to know the boundaries up front, and let them be innovative within those boundaries, than to pretend they don't exist. After all, this is not an academic exercise, but a real-life situation intended to produce a new solution to a problem that will actually be launched.

The best approach is to recognize and understand the boundaries—embrace them and make them part of the ideation process. I've seen more clever solutions generated by people who accept the boundaries, and innovate around them, than those who ignore the boundaries. Take for instance the innovation team from Bosch, who in 2004 was charged with introducing a new circular saw in North America where there was already an established market with a clear market leader. The overriding boundary given to this team was that the new saw would be adopted by both the user of the saw and the big box stores (such as Home Depot and Lowe's) yet be *priced at or below* that of the leading brand. No cost could be added, yet the saw had to attract significant attention! This is the business equivalent of the pharaoh telling the people to make bricks with no straw.

Yet, instead of being a creativity-killer, this boundary opened great opportunity for clever design and innovative thinking. One of the key issues of circular saw users was that it was difficult to stay on the cut line often resulting in an inaccurate cut. Given that the target audience for the saw included professional custom woodworkers where precision cuts were important, remaining on the cut line was critical to them. The team got busy on designing the best possible solution to this problem. While one of the cool early solutions created was a spectacular laser-sighted guide that would self-adjust or warn users if they were deviating from the cut line, it was a very expensive idea. Thus, the team had to look for other solutions that would add no cost.

They reviewed the contextual information to find out why staying on the cut path was so difficult for a professional woodworker. What they found is that as the saw moved forward, sawdust would be scattered by the motor's fan, often right onto the cut line. Now that they had a better understanding of why the problem existed, they redesigned the saw by turning the motor's fan *forward*, directing the fan to blow the sawdust off the cut line, thus making it visible at all times. In this example, the team used the source of the problem, the motor fan, to generate the solution, adding nothing to the unit cost. This new saw had many additional features that targeted important but unmet needs and became a major homerun for Bosch, with the saw winning the *Popular Mechanics* award the year it was released.

Over the two decades I've been working with innovation and creativity, I've seen this happen time and again. People can actually be *more* creative when they know the boundaries because they have to create a solution that will not only meet the objective, but also overcome these boundaries. Having teams generate ideas without fully understanding or considering the boundaries is an exercise in futility, since, at some point in the process, these boundaries will be imposed and the ideas will be short-lived if they do not stay within the boundaries. If, instead, the boundaries become part of the creativity exercise, then not only is a breakthrough solution possible, but it also has a strong chance of being implemented. Innovations that never see the light of day, after all is said and done, are not really innovations. They become relics in the Museum of Lost Ideas.

Using Resources to Solve Problems

One of the most important considerations when generating new solutions are the *resources* available throughout the ecosystem. Resources can consist of people, technology, items, platforms, and even things such as empty space, time, and money—virtually anything that is already present within the scenario. In the practice of TRIZ, the Russian theory of inventive problem solving, resources play a very important role in helping to discover new ways of solving a problem. Let's look at each of the primary resources in detail.

Time Resources

The resource of time can yield interesting insights into solving a problem. Using time resources may include accessing unused time, finding a way to multitask something, changing the order of execution of certain activities, and so on. For instance, for people without insurance who need to have expensive tests done, "unused time" creates a unique solution. The high cost of a diagnostic test is driven in large part by the significant cost of the equipment needed to conduct the test. Similar to airlines with unsold seats or hotels with unsold rooms, companies that own these testing facilities could offer deeply discounted rates for appointments scheduled during equipment downtime, for example, after normal business hours, when there are empty time slots on the schedule. Thus uninsured people get a good rate, although they may have to go at odd hours to get the test done, and the testing company makes money for time that would have yielded no revenue.

The example of shared medical appointments is an innovation that uses time resources by multitasking something that used to be done sequentially. Other types of time resource manipulation include pre-prepared versus real-time, automatic versus manual, flexible versus standard, and continuous versus periodic action. When looking at time as a resource, it is important to understand all the different aspects of when actions take place.

People Resources

In most social scenarios, people are at the heart of the solution. Whether it is improvement in the education system, saving natural resources, solving our health care problems, or improving economic development, it all comes down to the human players who need to conduct parts of the job. Some examples of innovative uses of people resources include using parents to volunteer time in the classroom to reduce the number of teachers needed and provide assistance and support to the growing number of students in the classroom. In the health care setting, less trained medical professionals are taking on more and more of the physicians' workload, allowing the physicians to focus primarily on diagnosis and medical advice. In the military, there is increasing reliance on members of the local community and tribal leaders in battle areas to become liaisons and provide support. The group Safe Passages in Chicago uses former gang members as well as recently retired veterans to serve as additional support to the police force to monitor and report violence or attempted violence on kids going to and from school.

This example also illustrates another TRIZ principle known as, "Make your enemy your friend." The goal is to basically take that which has caused the problem and use it as part of the solution. By using the gang members, or members of the tribal parties in Afghanistan, this principle is being played out; that which used to be part of the problem, gang members and Afghan fighters, are being used as a resource to solve the problem. This is a brilliant technique to use in all situations and can produce amazing results.

When looking at the social scenario, identify all of the people resources within the scenario going beyond those who have been identified as job executors, beneficiaries, and so on, as well as those who are the "standard" people within the platform. Some of the best uses of people resources are those where people on the periphery of the scenario or outside the traditional ecosystem are used.

Teach for America is an example of using nontraditional people resources. This program recruits recent college graduates and business professionals to spend two years teaching in low-income community schools. These new resources, the nontraditional teachers, bring new hope to areas of educational inequity. Teach for America has done an amazing job in meeting the related job needs of these nontraditional teachers, especially the recent college graduates. Its compensation program demonstrates the degree to which the organization acknowledges and addresses the needs of these teachers by going beyond the traditional salary and benefits, and including money to repay student loans, loan forbearance, paid interest for their tenure on staff, scholarships, benefits from graduate schools, and money for relocation. These are obviously attractive benefits to a recent college graduate. Has this strategy of ensuring satisfaction of the needs of the ecosystem members paid off? The numbers speak for themselves. The organization launched in 1990 with 500 of these nontraditional teachers and has grown to over 28,000 over a 21-year period. The alumni from this corps of teachers have included leaders from business, politics, entertainment, and professional athletes.

Space Resources

Space is often one of the least utilized resources in problem solving generally because it is just there; empty space is basically invisible to us and is therefore often overlooked. After Hurricane Katrina struck New Orleans leaving thousands without a home, the empty Superdome football stadium was used to provide temporary shelter for displaced people. In the United Kingdom, the Young Foundation has been advocating for the rights of communities to use land and buildings that are unused space. "Sharing the public estate with civil society is a way of getting more value from public assets and supporting community engagement at the same time. Examples include bringing more empty office blocks or unused green spaces into community use, or opening up school libraries and other public buildings which are locked up at night."[2] In the United States, Air Force bases that have been closed have been converted to public uses such as community college campuses, retirement centers, and prisons. In addition, many other uses have been found for the buildings and surrounding land.

Resource Inventory

Once the platform has been identified for the Ideation Strategy, it becomes imperative to develop a resource inventory. This should be done as a group exercise to ensure that it is as complete as possible. The inventory will be used extensively by the ideation teams as they work on developing solutions that will meet the needs of the ecosystem members.

Table 4-1 illustrates just a sample of resources that exist in the education scenario.

Resources are all around us. The trick is learning how to recognize them and apply them as an asset for addressing unmet needs. The most effective method I've seen for teaching people how to identify and apply resources to solve a problem is the Titanic Game, developed by Ellen Domb, Ph.D.,

Table 4-1 Resources Available from the Education Platform

Platform Element	Resources Available
Main infrastructure	School buildings, school administration, land that the buildings are on, multipurpose rooms, federal financial resources, etc.
Subsystems	Classrooms; individual campuses; to finance new programs and buildings; library; arts, choir, band rooms
People	Teachers, assistant principals, volunteers, janitorial staff, librarians, IT staff, parent volunteers, administration staff, etc.
Processes	Curriculum, education policies, requirements for graduation, teaching certification process, etc.
Technology	IT systems, video systems, campus security systems, etc.
Parent-specific additions	Financial resources, support for homework, time to give to the school, their home, etc.
Student-specific resources	Their focus, their time, backpacks, materials from home, etc.
Teacher-specific additions	Background knowledge and experience, personality, customization of curriculum, creative solutions for teaching the material, relationship with parents and children, etc.

an internationally renowned TRIZ expert. The focus of the game is to help people learn to see the resources that exist in the environment that are often overlooked. The game provides an experiential method of learning how to identify these resources and use them to create innovative solutions to a problem. Dr. Domb has graciously allowed the game to be included in this book with the hope that innovators in the social space will be able to apply this thinking to address the social challenges they struggle to solve.

The Titanic Game is often conducted with the innovation team during the kickoff of an ideation session. Team members have been exposed to the Ideation Strategy and are ready to generate ideas. I will often have the team spend time before the game identifying the resources within the scenario. Then after team members have played the game, I have them go back and see how many more resources they can identify. The number of additional resources found after the game is astounding.

The game can be successfully implemented by teams with as few as 10 people or with as many as 150. The participants are divided into groups with each group assigned to a part of a ship—one group has the kitchen, another the exterior of the ship, another the staterooms, another the dining room, and so on. The teams can use anything in their part of the ship to build their solution to save the passengers.

To start, the game is framed for the team members with the following six facts:

1. The ship has just hit the iceberg.
2. The engines are still running but will stop after an unknown amount of time.
3. The ship will sink in two hours, and the ship's officers know this.
4. The nearest rescue ship is four hours away.
5. There are enough lifeboats onboard for 1,178 people, and yet there are 2,224 people on the ship.
6. In the North Atlantic, a person in the water can live approximately four minutes.[3]

The object of the game is for the team to find a way to save all 2,224 people using only the resources that were available to the actual Titanic passengers. Many times the teams overlook the fact that there is a rescue boat that will be there in two hours, so they have to keep the people out of the freezing water for only a short time. This becomes an important learning point during the debriefing since whether they are trying to come up with a two-hour solution or a two-day solution makes a big difference.

Similar to reality, teams often set out to solve a problem with preconceived notions of what must be accomplished instead of confirming what indeed must be accomplished (this is why we insist on getting the stakeholder requirements up front—so we know what must be produced). In the Titanic Game, teams were more successful in generating solutions when they focused on keeping the people out of the water for only two hours.

I've conducted this game numerous times, and I'm always amazed at the solutions teams come up with to solve the problem. Teams in the dining room group often lash tables together with linens to make floating platforms; the team in the staterooms realize that if they throw a bunch of bed linens overboard near the gash in the hull, the force will suck the linens into the hole and potentially plug a good part of it to slow the sinking. The kitchen team will often use large cooking pots to hold small children, or even smear lard on the bodies of the people that would have to be in the water since this would delay the onset of hypothermia. I often find the most interesting solution to take place when the team decides to "make your enemy your friend"—the enemy being the iceberg. Some teams use the lifeboats as shuttles to take people to the iceberg where they could be wrapped in blankets and linens to wait for the rescue ship. Within 20–30 minutes, without fail, the teams have identified a significant set of resources aboard the ship and have created innovative solutions to the problem. Generally, once the game debriefing has taken place, people start asking the same question, "Why didn't the people on the Titanic do this?"

The answer is the same as that we see in business, government, and everyday life. In the midst of a crisis, people often do not see the options clearly, do not have time to put a plan together, and do not see the wealth of resources all around them. When a crisis hits in our work areas, whether business, nonprofit, or government, do we stop and ask, "What is the ideal final result that will solve the problem?" and, "What resources do we have right now at our disposal that could help?" Most of the time, we are not in a crisis situation, and we have plenty of time to plan and to put a strategy in place. We know that there will be another natural disaster (hurricanes, tornados, typhoons). We know there are hungry people everywhere. We know there are homeless people. We know that our education system is broken. We know that the health care system is unsustainable in its current form. What is the ideal final result? What resources do we have to solve the problem? What is stopping us from acting now instead of waiting for the crisis to hit? It is better to start working on these issues now.

The Process—"Let's Start Innovating!"

Once the investigation phase has been completed, it is time for the exciting part of the process—developing solutions. While some people approach this part with glee and excitement, others view it with dread. As we discussed earlier, not all people are comfortable with developing creative solutions. I would argue that all people are innately creative; the challenge is to give them the tools they need to release their creativity. The benefit of having the inputs we have obtained thus far is that the creativity becomes a problem-solving activity with a very clearly defined and understood problem.

Who Should Be Involved?

With problems such as those we are trying to tackle, the team involved in generating the solution should be composed of individuals from various sectors—public, private, nonprofit, and so forth—and should represent various functional areas within the organizations—marketing, program development, operations, field personnel, and the like.

In the case of a health care innovation for a *hospital system*, the team should include people from the physician side, some patients or people who represent patients and work with them daily, people from the nursing staff, the hospital administration, and any community outreach entities that are part of the team, to name a few. If instead the focus was on innovating the platforms of the *country's health care system*, then we'd want to include people from the state's department of insurance, the Department of Health and Human Services, from nonprofits who drive change in these areas, as well as patients, physicians, and hospital administration personnel. The goal is a diverse group that can look at the problem from various angles, challenge each other, and most importantly think beyond what exists today and build a vision of what could be.

The Structure of an Ideation Session

An ideation session is defined as a set of activities that take place to generate concepts that adhere to the guidelines of one Ideation Strategy Guide. Thus there is at least one session for each Strategy Guide and for each Strategy Guide there can be multiple sessions (these will be referred to as *ideation sessions* going forward).

A session can be conducted in many different formats: it can take place in a group setting over two to three days; it can be conducted asynchronously with virtual platforms; it can be conducted over a period of time with a combination of both group session and asynchronously. The key is that the session is defined by the guide and not by a point in time. I have found the most effective method to be conducting one in-person group session to start the process and get the ideas and concepts going. Then initiate an online (asynchronous) forum for two to three weeks of further idea generation and refinement, followed by a group session to have concept review and final convergence of the concepts into concept roadmaps.

Within any part of the ideation session, the process should include a combination of rounds of divergent (broad, expansive) and convergent thinking (narrowing down, selecting). Traditional brainstorming sessions follow this pattern naturally where the team begins with divergent thinking to generate new ideas, build on ideas, and explore many different ways of solving the problem (a divergent process). Then, the team will begin to prioritize and consolidate the ideas, filtering them down to a small number of winners (the convergent phase). However, it usually only goes through this cycle once.

Most successful innovation takes several cycles of divergent and convergent thinking. The danger with stopping after the first divergent phase is that often these are the top-of-mind ideas that may be good, but generally are not the "best" idea. The best ideas are those that are generated after the initial set is out on the table and have gone through a second round of divergent thinking, thus ensuring that the ideas have been stretched farther than the obvious.

Six-Step Session

During the last 20 years of working with ideation sessions, I've found that there are six steps to be executed whenever a team is trying to develop new solutions. With these six steps you can choose from a variety of creative thinking tools and techniques. It is the framework that will ensure the pattern of divergence and convergence. The tools used for each of the steps can be customized and varied, although I've included my favorite tools in each step as an example. Having a solid process for ideation removes the mystique, engages the participants more fully, and makes the process simple to facilitate.

The six basic steps are as follows (*Note*: this assumes that you have already conducted the investigate step and have the context already in hand):

1. *Review the needs and context:* ensure that the team has a clear understanding of the opportunity, the context, and any other information about the need.

2. *Ideation Round 1:* (a) Get the usual suspects out on the table using a divergent thinking tool; (b) pull initial ideas into groups of similar ideas or initial concepts (round 1 ideas—these are generally still at the sticky-note size of idea and not full-blown concepts).

3. *Ideation Round 2:* Engage in lateral thinking techniques (divergent phase) to either expand on the first set of ideas or to generate new ones. Ideation techniques traditionally have some sort of stimulus, whether it is a tool, a story, a prop, or a framework—something that helps get the thinking started and guide the thinking away from that which is expected. Once the divergent phase is complete, then begin to pull together the ideas into preliminary concepts (convergence phase) (round 2 concepts—these are becoming concepts instead of simple ideas).

4. *Ideation Round 3 (optional):* Go divergent again to further push the ideas to the extreme; converge to round 3 concepts (more comprehensive concepts).

5. *Formulate final concepts:* Pull from all rounds of concepts and create the final concepts on a concept board.

6. *Assessment:* Test the concept against the needs of the ecosystem members as well as the stakeholder criteria to ensure that they are on target.

As mentioned earlier, within each step there are various tools and techniques to use. For instance, many organizations have favorite tools for getting the initial set of ideas flowing. Some have extensive lateral thinking tools and exercises. Others have detailed assessment tools. This book contains several examples of what I've used successfully in the past; additional tools, and detailed instructions on how to use the tools discussed here are available for download at our website: www.TheSocialInnovationImperative.com.

Tools are an important means of engaging the entire group, building solid consensus, strengthening teamwork, and ensuring that everyone participates. Tools help to democratize the group, removing barriers of political hierarchy. Good tools also help everyone become more creative by providing visual, tactile, and experiential devices, analogies from other industries and scenarios, and triggers to get people to break out of their preconceived constraints and frameworks.

For this chapter, I use a case study that most people can relate to. The innovation challenge is to prevent obesity in pets. During extensive research with pet owners and veterinarians, this was identified as a key issue. One of the Hi-Impact Constraints was that in multi-pet households, the pet that needed to lose weight would not eat the diet food it was supposed to eat. It also tended to eat too much because it would access the food from the other pets' bowls in addition to its own. This is a big problem when the owners paid for specialty food for the pet that needed to lose weight.

Round 1: The Usual Suspects

Most people involved in the ideation sessions will be excited to begin since they've been exposed to the Hi-Op Needs, and have worked through the context, which probably has the problem solving wheels going at full speed. It's hard to stop people from generating ideas with this influx of great information about the ecosystem members. So the first step is helping them get those top-of-mind ideas flowing and out on paper. The "pass card" tool (see Figure 4-1) was developed by an innovation and creativity expert Terry Richey, founder of *The Next Idea* and author of *The Marketer's Visual Toolkit*, and I've found it to be one of the most effective tools for generating many ideas from the team in a very short period of time, then prioritize them down to a handful. Since each participant has a card, there will be three ideas generated for each person for a total of three times the participant's ideas and times the number of participants' prioritized ideas. Thus, if there are 20 participants, this 15-minute exercise will generate 60 ideas which are filtered down to 20.

All of the ideas remaining on each Pass Card at the end of the exercise are transferred to an *Idea Note* (usually a 3 × 3 Post-it note) so that the ideas can be worked with more easily. A sample idea note is shown in Figure 4-2.

A quick definition check: throughout this book, *ideas* are thoughts that will fit on a typical 3 × 3 Post-it note; a *concept* is a combination of ideas that have been aggregated, contains features, and has a detailed description.

These remaining 20 ideas are then put into a short convergence phase where they are grouped together into preliminary concepts or simply aggregated into affinity groups of like ideas (often revolving around

Figure 4-1   Pass Cards

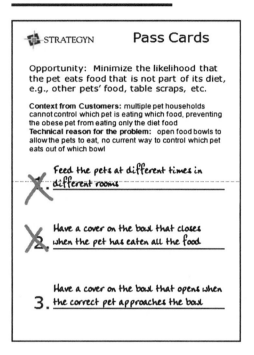

platform), yielding the "Round 1 Ideas" as shown in Figure 4-3. In our example, there are some ideas that deal with the physical properties of the bowl including a lid on the bowl, a scale on a pad under the bowl, and so on. Another set of ideas is focused on programs that could be implemented such as a weight watcher type program or a vet discount program when the pet loses weight.

Round 2: Divergent Thinking Techniques

Once the traditional ideas have been gathered, it is time to expand the direction of the thinking into new possibilities—go divergent. Edward de Bono introduced the concept of lateral thinking in a book by the same name published in 1970. He cites lateral thinking as a *deliberate process* that the mind uses to change the existing patterns that have been created by our highly efficient brains. Because our brains are so good at creating patterns and affiliations, it is necessary to break free of them.

Figure 4-2 Pass card ideas transferred to idea note

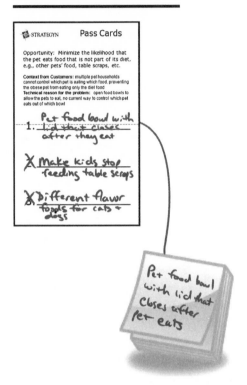

Lateral thinking techniques take our minds to a new level of thinking and exploration, to a level that is unexpected, nonlinear. Lateral thinking is a tool for creating insight, for breaking through patterns that our minds have created to be effective but that often prohibit us from being creative. It is a process, a method, that can be learned and taught in order to help people

Figure 4-3 Round 1 ideas grouped

New type of eating device

Programs to encourage weight loss

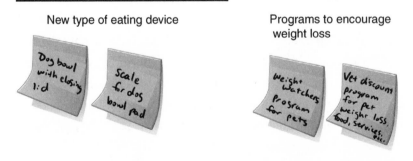

become more creative. "Lateral thinking is concerned with breaking out of the concept prisons of old ideas. . . . Liberation from old ideas and the stimulation of new ones are the twin aspects of lateral thinking."[4]

Over the past 40 years since de Bono introduced this concept, there have been numerous lateral thinking tools developed to help stimulate thinking and guide creativity. The key to lateral thinking techniques is to provide stimuli to the participants so that they can think about the problem from a new perspective. Without these stimuli, teams will find that they are generating the same ideas as before.

In most cases, the ideas generated from the first round will be traditional ideas, those that fit in the existing patterns of the world. This happens because the participants are simply asked to provide an idea that will solve the problem. There is not a stimulus provided to get them to think in a new dimension.

There are a number of different methodologies and theories on problem solving, but one that I find to be most foundational with principles used in many other methods is TRIZ, the theory of inventive problem solving. This internationally acclaimed process was founded by Genrich Altshuller from the former Soviet Union after World War II. Altshuller scoured thousands of patents from around the world to understand what made these inventions unique and what techniques they applied to solving problems that customers had or limitations of existing products. During this process he discovered that there were many common patterns that explained the way the invention was able to solve a specific problem and become a marketable solution.[5]

While much of the TRIZ methodology was designed to address problem solving with physical products, the principles that were identified have been adapted over time to creative thinking stimuli that can be applied to any situation.

In my work with Terry Richey of The Next Idea[6] and in subsequent development of creativity tools for clients with unique market challenges, my colleagues and I created a series of tools that use many of the TRIZ principles in a way that is fun, engaging, and highly thought provoking. The tools provide a framework that allow innovation teams to break through their existing thinking patterns. Within the tool, there are modifications of the

TRIZ principles—what we call Creativity Triggers. Thus, for the remainder of this chapter, I reference Creativity Triggers' version of the principles as opposed to the TRIZ version. Table 4-2 illustrates a few examples of the difference between the two methods.

Table 4-2 TRIZ Principles and Creativity Triggers and their Examples

TRIZ Principle	Creativity Triggers	Example
Extraction: Extract the "disturbing" part or property from an object; extract only the necessary part or property from an object	*Remove it:* Remove an input or output of the system. Hide something in the solution or the environment. Dissolve a feature after its useful function	*Product:* Biodegradable stitches that dissolve after a period of time *Program:* Insurance company that offers disappearing deductible where the deductible is reduced every year that the customer does not have an accident
Consolidation: Consolidate in space homogenous objects or objects destined for contiguous operations; consolidate in time contiguous operations	*Combine it:* Combine one or more processes, programs, services, teams, etc.	*Product:* What Off! combined processes when it created its bug spray with sunscreen *Program:* San Diego's one-stop center for victims of domestic abuse containing police, prosecutors, social services, and nonprofit support groups all in one location
Periodic action: Replace a continuous action with a periodic one; if the action is already periodic, change its frequency; use pauses between impulses to provide additional action	*Continuous/periodic action:* Make a periodic process or activity continuous or make a continuous activity periodic	*Product:* Impulse water sprinkler produces droplets resulting in less water damage to the soil *Program:* Year-round school programs reduce the length of time that children are out of school at one time

Random Stimulus Tools

One of the Creativity Triggers we use is a lateral thinking technique that uses random stimuli. By pairing unusual items and forcing strange combinations, interesting ideas are generated. Randomly assigning a stimulus (trigger) to a problem can create unusual combinations, which I've found can often generate some of the most innovative solutions. Let's look at a few of the executions of these stimuli in practice.

One simple but effective execution of the triggers is a set of Idea Cards that I've created based on the TRIZ and Creativity Trigger principles. Each card has a trigger with a description and an example of how that trigger has been used in reality. The triggers are grouped by a set of categories, or divergent paths, designed by my colleague Lance Bettencourt, Ph.D. and author of *Service Innovation.*

How to play. Keep in mind that these cards are used after the first set of ideas has been generated in Round 1, and the purpose is to expand on those ideas or generate different, more innovative ideas. The team members randomly select a card and try to apply the trigger on the card to either the existing ideas (as shown in Figure 4-4), to the Hi-Op Need statement, or even to one of the issues identified in the context.

Figure 4-4 Using a Creative Idea Card to Stimulate New Thinking

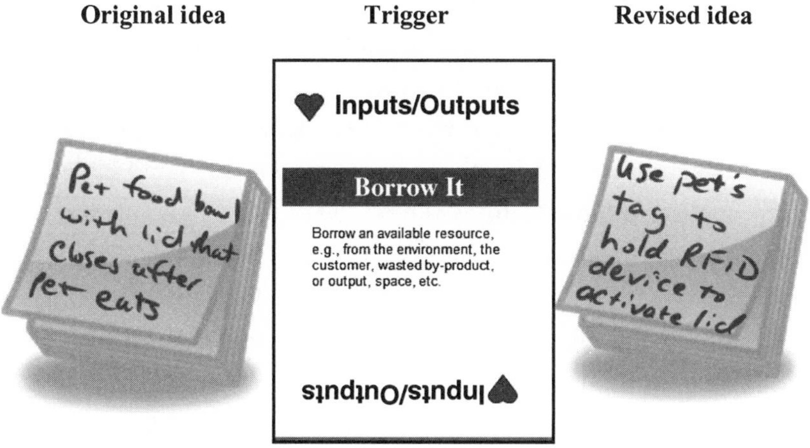
Original idea
Trigger
Revised idea
Pet food bowl with lid that closes after pet eats
Inputs/Outputs
Borrow It
Borrow an available resource, e.g., from the environment, the customer, wasted by-product, or output, space, etc.
Inputs/Outputs
Use pet's tag to hold RFID device to activate lid

Figure 4-5 Idea Cards

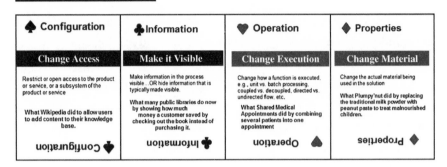

Returning to our example of obese pets, the teams began formulating ideas around a new pet food bowl that would have some sort of lid that prevented access to food. One problem they hadn't figured out for this idea was how to make sure that the right pet ate the right food. During the Idea Card exercise, they drew the "borrow it" card which says, "Borrow an available resource, e.g., from the environment, the customer, wasted by-product or output, space, etc." This prompted the idea to include an RFID (radio-frequency identification) device in the pets' tag or their collar which would be tuned to their bowl only. This would allow only the pet with the matching tag to access the food. Some other examples of the Idea Cards are shown in Figure 4-5.

Another good tool in the random stimulus category is the Idea Transformation Grid (see Figure 4-6). This grid is used when trying to transform a set of features into something more innovative. For instance, if an existing platform is being used, and the team is trying to develop new features for the platform, the features can be listed and Creativity Triggers applied to those features. The value of this tool is that the triggers are applied to very specific items. The tool can be used to transform features, issues within the context, existing ideas, traits or characteristics of a scenario, and so on. Going back to our example, the features listed might include things like the bowl, the bag of food, the measuring cup

Figure 4-6 Feature Transformation Grid

Features	Remove it Remove a key aspect of the feature, process, or activity . . .	Make it visible Provide the customer with more insight into what is happening at different stages . . .	Move it Move a part, people, process, or any other component in the scenario . . .	Change it Significantly change how the feature works	Combine it Combine the feature with something else–another process, product, service, etc.	Rethink it Start a whole new movement with the scenario or a feature

used to scoop the food, the owner, and the pet. All these are features that can be transformed. Then the triggers are applied to the features. I strongly urge my teams *not* to go in order (people tend to want to start in the upper left corner and work their way across and then down). This is not effective since there is probably not enough time to fill in all the cells, and people will tend to work on the combinations that make the most sense. For instance, they will gravitate toward combining it with the bowl feature since that will be easy to work on. Instead, I have the teams roll a pair of dice. One die tells which way to go across, the other determines how many rows to go down. So, if a 5 and 4 are rolled, participants would take the fifth feature and apply the fourth trigger (or vice versa). The key is to make it a random pairing. So, with the 5 and 4 roll, the team will work on developing ideas for that combination for 5–10 minutes or until team members can't think of anything else, jot down the ideas in the cell, and then they roll again repeating the process.

The cards and the grid require very little customization; they can be reused and applied to numerous ideation sessions.

Analogies and Case Examples

Many creativity gurus emphasize the use of analogies, scenarios, case examples, and stories to get people to think about the problem by viewing it from a different perspective, a different industry, a different personality. These are highly effective tools for creating new ideas. Two of my favorites are scenario boards and idea boxes.

Scenario boards, such as that shown in Figure 4-7, tell a story about a person in the target audience for whom we are trying to generate a solution.

Figure 4-7 Scenario Boards

Ideation Scenario Board

Owner Scenario – Dog

Jerry Anderson

Age: 45
Married with two teenage children. Lives in the suburbs of Denver.
Has three dogs and is a quality loyalist.

I have always had dogs, as a kid, as a teenager, and as an adult. That's dogs with an 's'— lots of dogs! Our family has always considered the dogs an important aspect of the family unit. But, like teenagers, dogs can be pretty unpredictable. Especially when something changes around here.

Questions for Value Creation

1. What solution can be developed to help Jerry get Rudy's aggression under control?

2. What services could be created to help owners manage multi-pet households with territorial or other types of aggression?

3. How can Jerry's vet help this family maintain peace?

Lately, I've been seeing so much more aggression from Rudy. He's the one at the far left in the picture. He's been finding sport lately in picking on Sam (the one at the far right — there's a reason they're not next to each other). I've noticed the aggression coming out when they eat, if one of them has a toy the other wants it; just about anything can tip them off. Once it starts, then they all tend to want to jump into the fray. I'm worried that one or more is going to be seriously hurt if I don't get this behavior under control. Not to mention the kids if they try to break up the fights.

The boards outline the issues the person is facing (which are based on the Hi-Op Needs and/or Hi-Impact Constraints) and pose questions at the end.

Idea boxes (see Figure 4-8) are another way to trigger new ideas by illustrating how other industries are solving problems similar to the one the team is trying to solve. In this case, solutions concerning human weight control are presented as stimulus for animal weight control, as well as the Nike+ running program that helps runners keep track of their progress.

Note that these two tools take time, effort, and a bit of creativity to develop and must be customized for each ideation session based on the audience, the needs, and the context. While they may take time and effort

Figure 4-8 Idea Box Cards

C O N T R O L

From Intentions to Reality
Stay the course. Make it automatic.

What possibilities emerge from this innovative idea that helps dieters stay in control, manage portion sizes, and all-in-one kits? Can these concepts be used to help pets get to and stay at a healthy weight?

Creating a lifestyle from good advice.
The STAX system for organizing meals is the missing link between diet books and nutrition programs and the lifestyle that is required to lose weight. The STAX cooler ensures that the dieter has the right food available when they need it. The custom timer schedules meals to ensure that none are ever missed, keeping the dieter on track all day long. The system of color-coded portion control containers ensures that the dieter always knows exactly how much to eat. What, when, and how much – it's all covered!

(continued)

Figure 4-8 *(Continued)*

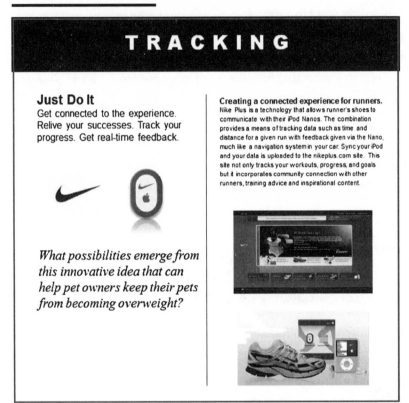

to produce, they are well worth it. They yield tremendous results and really get the team members to think about how they can apply solutions that others have successfully implemented (in the case of the idea boxes) and to put themselves in the shoes of the target audience with the scenario boards.

Positive Deviance Techniques

The theory of "positive deviance" was developed by Jerry Sternin in his quest to find "outsider solutions to local problems." In his groundbreaking work on malnourished children in Vietnam, he noticed that some of the children in the same impoverished area were actually quite healthy (they deviated from the norm of the area but in a positive way). He studied these outliers to find out what their mothers were doing differently and identified several solutions they used including adding small amounts of shrimp and crabs to the children's meals, adding greens from sweet potatoes, and giving them more frequent but smaller meals. By having the mothers of these positive deviants train the other mothers in the area, the innovative solution rapidly spread throughout the villages.[7] This is a great example of using people resources that already exist in the area. This same technique has been used to identify classrooms with exceptionally good performance to find out what the teacher is doing differently.

These techniques are but a handful of the ones that exist and only a handful of the ones that I use during a typical set of ideation sessions. I hope they provide good insights concerning the value of lateral thinking and the benefits of using such techniques to stimulate creativity for those of us who don't have those pure "out-of-the-box" genes. With the right stimulus, we can all be very creative.

Formulate Concepts

Once divergent thinking has taken place, the ideas are ready to pull together into concepts. Often, several ideas are combined into a single concept with the individual ideas serving as the features. The concept itself needs to have enough detail to determine whether it will meet the needs of the ecosystem members whose needs are being addressed and any constraints that have been overcome. As the concept is formalized, there are often notes that need to be added for potential issues that team members have identified, areas that may require more exploration, questions that need to be answered, and so on. It is also where the features get matched to the Hi-Op Needs and/or Hi-Impact Constraints to ensure that the concept and features are still focused on addressing the needs. It is very easy for an idea to take on a life of its own and generate excitement among team members. Before long the concept has strayed from the original intended goal and no longer addresses the need. This happens a lot which is the reason for the feature-need section on the concept board. Features that have no connection to Hi-Op Needs should be carefully weighed to ensure that they are worth the cost, the effort to develop, and so forth. In some cases these types of features are added for aesthetics or overall appeal of the concept on an emotional level (i.e., the "cool" factor). It is important to weigh the benefit of these features relative to their cost, development time, and the like.

There are likely to be several convergence phases, depending on how many rounds of divergent thinking are planned. As the ideas begin to converge into concepts, it is important to apply more divergent thinking to them to see if the ideas have been stretched far enough. Can anything be added to the concept to address a Hi-Op Need of another ecosystem member or to overcome a constraint?

As the concepts are generated, it is important to gauge the progress being made with respect to the Ideation Strategy Guide. This checking-in point should take place after every convergence session to monitor progress and ensure that the ideas or concepts are well on target. It will also help to identify any areas that are weak in meeting the objectives of the Ideation Guide. For instance, if the guide is looking for short-term concepts that will focus on a given member of the ecosystem, has that been delivered? Have all the needs that were targeted been addressed with one or more concepts? The Post-it notes can be placed on the grid to identify where the ideas lie within the dimensions of Development Effort and Impact on the Ecosystem (see Figure 4-9).

Figure 4-9 Progress grid for ideas

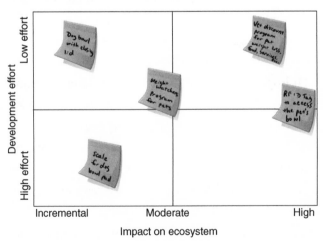

Assessment

The assessment phase of the ideation session takes place at two different times. The initial "quick look" assessment can take place at the end of the in-person group ideation session. This assessment is rather informal and is based on the teams' perspective of how well the concepts meet the needs, constraints, stakeholder criteria, and so on. A simple matrix is used with either +, 0, or −, or it can be done with three colored dots (green, yellow, and red) as shown in Figure 4-10, which is from a photo from a recent session. The goal is to get a quick visual of which concepts are tracking as the top ones to pursue and indicate where the concepts are having problems. The criteria on the worksheet is based on the Ideation Strategy Guide and are customized for each session.

Once the concepts have been fleshed out and detailed, then a more formal assessment takes place. This assessment is a comprehensive overview of which jobs the concepts address, for which ecosystem members, and which constraints it overcomes. It is basically a Pugh matrix where the

Figure 4-10 Quick Evaluation Matrix

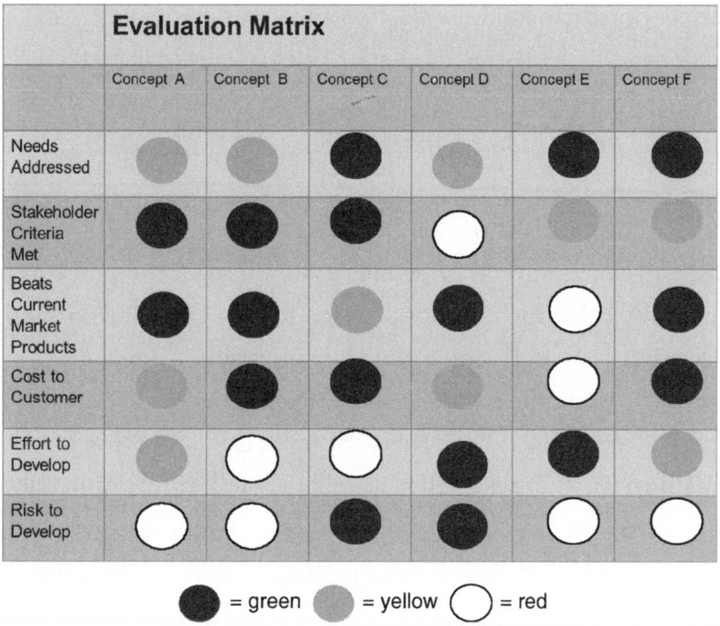

Evaluation Matrix	Concept A	Concept B	Concept C	Concept D	Concept E	Concept F
Needs Addressed	yellow	yellow	green	yellow	green	green
Stakeholder Criteria Met	green	green	green	red	yellow	yellow
Beats Current Market Products	green	green	yellow	green	red	green
Cost to Customer	yellow	green	green	yellow	red	green
Effort to Develop	yellow	red	red	green	green	yellow
Risk to Develop	red	red	green	green	red	red

● = green ● = yellow ○ = red

concepts are evaluated with respect to the needs, constraints, and stakeholder criteria. (See Table 4-3.)

Platforms for Collaboration

Creating solutions for the complex problems posed by social scenarios requires the efforts of many people from various sectors. As cited in the *Stanford Social Innovation Review (SSIR)*, "Social innovation increasingly requires collaboration among diverse networks of nonprofits, government agencies, corporations and private citizens. These networks promise a wider range of ideas, better use of resources and faster solutions than do traditional monolithic entities."[8] Many in the social innovation space believe that true breakthroughs in social innovation will not take place within any one organization, no matter how large, but that it will happen at the intersection of the sectors, where they meet. Another article in *SSIR*, "Collective Impact," cited the following example to support this point:

> The scale and complexity of the US public education system has thwarted attempted reform for decades. Major funders, such as the Annenberg Foundation, Ford Foundation and Pew Charitable Trusts have abandoned their efforts in frustration after acknowledging their lack of progress. Once the global leader . . . the country now ranks 18th among the top 24 industrialized nations.[9]

It is clear that these problems require collaboration. But with such a large, global, and diverse group of people trying to solve the problems, how do we actually bring these minds together to generate breakthrough ideas? It is clear that, "To foster innovation, organizations need to develop places where they can come together and work creatively," this place is a *platform for collaboration*. The authors list three different types of collaboration platforms that are needed (actually they follow the same lines of the three sections of this book): exploration (investigation), experimentation (ideation), and execution (implementation).[10] A series of platforms

Table 4-3 Evaluation Matrix—Detailed Version

Hi-Op Jobs	Current product sat Product A—Baseline	Competitors' product satisfaction	New concepts Concept A	Concept B	Concept C	Concept D
Detect any disease or illness at its earliest possible stage						
Know with certainty that your body is free from disease, infection, tumors						
Reduce the risk of developing age-related diseases or disabilities						
Maintain the desired level of physical performance—stamina						
Get the proper amount of sleep						
Determine if any internal body parts (organs, brain, glands, etc.) are functioning improperly						
Maintain the desired level of muscle tone						
Stakeholder prioritized criteria			Concept A	Concept B	Concept C	Concept D
Development cost						
Time to launch						
Impact on scenario (i.e., likely to be adopted)						
Total scores						

that can provide this level of collaboration—from the beginning of problem investigation through the implementation or diffusion of innovations—can become the catalyst for true social innovation, removing the most significant barrier to date which has been an inability to get the sectors to work together in a common framework. Once the sectors—corporations, nonprofits, government agencies, and private citizens—have a means by which ideas and thoughts can be shared and built upon, social innovation will reach its tipping point, and we will see real movement take place.

Platforms for Exploration or Investigation

Platforms for exploration or investigation are designed to bring together diverse stakeholders to develop a common understanding of the problem. Following the methodology introduced in this book, the data supporting the understanding of the problem are quite substantial. A platform for collaboration that could bring together all the vital information uncovered by the methodology—the framing information, member needs, opportunity spectrum data, and context—would be vital for stakeholders involved in solving the problem in order to become fully immersed in the problem.

A key aspect of the investigation platform is the aggregation of problem framing, needs data, constraint data, and context data. These qualitative statements and problem definitions enable organizations to be ready to generate ideas immediately. Social innovation has the potential advantage of a true "open innovation network" when the information gathered in the first several steps (problem definition through context) is added to a global set of needs for the social scenario and the data made available to those who would help to solve the problem. My colleagues and I have been working to create what we call the DNA (discovery needs archive) of several scenarios including health care, education, and resource conservation. Most of my clients have been very willing to contribute this part of their studies to the greater good after a brief moratorium in which they can use the data for themselves.

How does this work? If you think about it, all school systems are struggling with the same problems; it doesn't make much sense to gather data for one school and then start over and do qualitative interviews again for another school. Instead the needs become part of the DNA. It is the quantifying of the needs in the new area that accounts for the differences in the population.

To demonstrate how this works, take an example in the health care space. We conducted a study for the California Health Care Foundation in 2006, another study for the Harvard/Kennedy School of Government in 2004, and another study for corporate client DSM. Through these studies we were able to assemble a sizable health care DNA database. Our health care DNA currently contains the needs and context of patients, providers, and employers; we hope to add hospital administration and government agencies to the DNA as well. We were able to take this qualitative data (the needs and constraints) and requantify them for a hospital system in the eastern United States. The reuse of these need statements saved the hospital system significant dollars and time. The team was able to immediately begin quantifying the needs, analyzing the market, and generating solutions.

Future benefits of the DNA database will include benchmarking and, in noncompetitive areas such as government, a sharing of the solutions created for each need. Imagine a platform in which the needs of the citizens relative to their city government are captured in a DNA. As governments work to solve these problems, their solutions are added to the DNA. Thus if the city of Austin comes up with a great solution for the need of providing safe areas for alternative transportation, such as bikes, pedestrians, and so on, any other city that finds that to be a Hi-Op Need can see what was done in Austin and borrow the portions of the solution which might be applicable.

This sharing of the investigation stage through an open platform will contribute substantially to the advancement of innovations in the social space, giving organizations working in these areas a substantial head start in the process.

Ideation Platforms

Ideation platforms are becoming increasingly popular, thus allowing teams to broaden ideation to include people external to their company, to people from other sectors, and to people around the world in order to include the best experts in the areas of unmet needs. This is also referred to as *open innovation* and can be extremely beneficial to a particularly challenging unmet need, or simply when the goal is to engage people from many different sectors in the ideation process. The systems are designed to accept ideas, route them to others in the network to contribute new thoughts and help overcome obstacles, allow peers to rate the ideas according to the ability of the idea to meet the needs of the members of the ecosystem, and much more. Such platforms allow individuals to participate in solution development from across the country and around the world.

Most systems make it possible for members to submit an idea; to collaborate online with others to build on an idea, add sketches, photos, and other rich content; and to do a quick assessment of the concept as it is being built. Two of these types of asynchronous ideation methods are Imaginatik and InnoCentive. InnoCentive has an added benefit of having a network of solvers that become additional resources to help solve the unmet need. There are also a few public idea sharing sites that are valuable for "crowdsourcing" or getting input from the general public. One such site is ManageMyIdeas. com, a Facebook type system for sharing public ideas. Microsoft has a platform for open innovation on its SharePoint 2010 software that allows ideas to be posted, managed, and assessed. The Social Innovation Exchange is a global platform that helps to facilitate the exchange of ideas, knowledge, and best practices across sectors and countries.

With the focus on innovation in the commercial sector, there is an abundance of choices when it comes to idea management software. It is best to find a solution that will allow a significant amount of information to be posted as the problem statement so that the Hi-Op Need can be listed along with the context information and stakeholder criteria; this way, anyone who participates in the web-based session will have all the critical information needed to generate an idea that will hit the target.

Experimentation Platforms

In "Platforms for Collaboration," the author submits that, "Most nonprofits and government agencies skip experimentation. Consequently, many social innovations go more or less directly from idea to implementation. Yet as social innovations cross boundaries and increase in complexity, experimentation will become the cornerstone of effective problem solving."[11] One major benefit of working with a needs-first approach like the one introduced here is that the experimentation can be done virtually. The assessment phase we discuss earlier becomes the experimentation where the team can determine how well the concept meets the stated needs of the target audience. This virtual lab will help to prioritize the concepts and make sure they have features that will have meaning for the audience.

The concepts that do well in the virtual lab testing are ready for the next stage of the experimentation platform—incubation. "Experimentation platforms give organizations a neutral environment for building and testing solutions in the simulated or 'near-real-world' contexts."[12] One of these platforms is operated by Business Innovation Factory (BIF) which creates "experience labs" for its clients. These labs allow the organization to fully test its ideas before implementation. In this platform the teams execute a process similar to the ideation methods we just went through where they create concepts that address the needs of the members within the scenario. Once these concepts are generated on paper, they are then molded into real-life concepts using prototypes, mock-ups, and simulations to see how the concepts will work in real life. This platform allows the innovation teams to continue working on the concept until they iron out all the wrinkles; it is at this point that the team will have the confidence and supporting data to develop a launch plan.

A new center in Austin, Texas, called Southwest Key's Social Enterprise Initiative is an experimentation and incubation platform that focuses on the social innovation space. "The center now serves as a launching point for community initiatives focusing on job creation, education services, social enterprise, and community empowerment."[13]

An important feature in these types of platforms is strong diversity among the people participating in the creation of the prototype or pilot program. This is especially important when dealing with numerous ecosystem members where adequate representation for each of the members and diversity within the member groups is vital. For example, if physicians are an ecosystem member, include both a Primary Care Physician and specialist, a private practice and a staff physician, a younger physician and older physician, and so on. The goal is to bring together diverse experiences and perspectives.

It is also imperative that these systems define a common set of success metrics that help different partners who are rapidly generating and testing new ideas. We have the benefit of having the Hi-Op Needs that are defined in the Ideation Strategy Guide to provide a significant level of detail on the metrics that must be hit. For example, we have the importance, satisfaction, opportunity percent, and context for the Hi-Op Need. We also have any jobs that are related to the Hi-Op Need (via the theme) and synergistic/conflicting jobs within the ecosystem for that need. This is an extraordinary amount of information to sort through in evaluating potential ideas.

In Summary

While the needs-first approach requires up-front work for identifying the critical needs, Hi-Impact Constraints, and detailed understanding of the problem, such an approach is significantly more effective at producing results. Organizations in the commercial sector using such approaches generate three to five times greater success rates than an ideas-first approach (depending on which methodology is being used and who is quoting the number). By generating an ideation portfolio, teams within organizations, and across organizations, have a clear road map for where they are going, what goal they need to hit and when. It allows the teams to *focus* their creative energy directly on the critical components of the social issue; such focus of creative talent produces results.

Articulating the path to success is key to gaining the momentum of the team. When my adolescent son struggles with a big problem that he is trying to solve and starts to get frustrated, we have a code to help him refocus. I ask him, "How do you eat an elephant?" and inevitably he will roll his eyes and respond, "One bite at a time." This friendly dialogue is a way to remind him that in order to tackle a problem that seems insurmountable, you must approach it one bite at a time, or in this case, one need at a time or one innovation at a time. The road map shows the team where the goal is, and the Ideation Strategy Guides help team members know what to do to accomplish each "bite" on the way. Armed with the tools and techniques provided, they soon will be on their way to tackling the elephant—the wicked problem.

Chapter | 5

DEVELOP A BUSINESS MODEL

One of the more difficult parts of social innovation comes down to who will fund the development of these great ideas, how will they be priced, and who will purchase them. Is a disruptive business model needed (as well as a disruptive platform), or can the innovation coexist within the platforms and business models of today? Can the job beneficiary afford the solution or will it have to be subsidized by another party? These are all questions that must be answered next in our quest for social innovation. In our vernacular, a concept's business model describes how the concept will create, deliver, and capture value and how it then gives back to the world. By using a concept, or consuming a product, a reaction takes place in people, their environment, their community, and the earth.

In July 2008, Bill Gates wrote a thought-provoking article in *Time* magazine called "Creative Capitalism." In the article, Gates argues that while capitalism has improved the lives of literally billions of people around the world, there are still "a great many that do not benefit because they have *needs that are not expressed in a way that matters to markets* [emphasis added]." While the needs of these individuals are often deferred to nonprofits and government entities, Gates argues that it is *corporations* that are best equipped to develop innovations and make them available around the world. An illustration he provides of mobilizing corporations to do good is the RED campaign in which hundreds of corporations agreed to contribute a percent of revenue to the fight against AIDS in Africa. This type of creative capitalism met the needs of all involved—valued research dollars were generated for AIDS research, corporations received PR benefit and increased customer traffic, and customers felt good about buying a pair of shoes or a shirt, knowing that some of their money was going to a good cause.

As Gates points out in the article:

> There are two great forces of human nature—self-interest and caring for others. Capitalism harnesses self-interest in a helpful and sustainable way, but only on behalf of those who can pay. Government aid and philanthropy channel our caring for those who can't pay. And the world will make lasting progress on the big inequities that remain . . . only if governments and nonprofits do their part by giving more aid and more effective aid.[1]

However, as the world continues through the great recession and governments themselves are stretched for funds to the point that they are cutting serious flesh from an already stretched system, it is obvious that this approach is not enough—we cannot put the entire burden on governments and nonprofits. We must somehow harness the corporate will to engage more fully in creating social value.

Creating Shared Value

Fast forward to January 2011 and a new concept called "creating shared value" has emerged. Introduced by Michael Porter and Mark Kramer in the January/February 2011 *Harvard Business Review*, this concept takes organizations well beyond corporate social responsibility and social entre-preneurship. The premise is that if companies were to put society's needs at the core of their strategy and operations, the companies would be able to identify new avenues of growth and gain an edge in becoming a sustainable organization:

> Companies must take the lead in bringing business and society back together. . . . [While] we still lack an overall framework for guiding these efforts . . . the solution lies in . . . creating economic value in a way that also creates value for *society by addressing its needs and challenges* [emphasis added]. Businesses must reconnect company success with social progress. Shared value is not social responsibility, philanthropy or even sustainability, but a new way to achieve economic success. It is not on the margin of what companies do but at the center. We believe it can give rise to the next major transformation of business thinking.[2]

In order to make this work, it becomes imperative that corporations understand the needs of not only their direct customers but also of the society in which their customers live, their communities, the environment, and so on. It will require a "far deeper appreciation of societal needs, a greater understanding of the true bases of company productivity and the ability to collaborate across profit/nonprofit boundaries."[3] Getting corporations this involved in the social innovation will catalyze the movement to new levels.

Why is so much pressure on corporations to come through with the creation of shared value? Because corporations have control over the entire production of their goods and services, how they are created, the raw materials used, the way products are brought to market, how they are priced, and so on. Once corporations begin to accept responsibility for the environment from which they attain raw materials, the society in which their products and services are used, and the community and context in which the customer operates, only then will sustained value generation be possible. Shared value can be created only when companies generate economic value for themselves in a way that simultaneously produces value for society by addressing social and environmental challenges. It is the organizations that understand the needs of society and integrate those needs into their business plan that will succeed moving forward. These organizations will become an integral part of society, not profit machines that are seen as the enemy by the environmental groups and distrusted by their own customer base.

In working through the issues identified in the article, I've developed a Shared Value Framework (see Figure 5-1) that can help organizations articulate the various parts of the framework and the interconnections between them. There are three sections—the company (including its employees and the environment in which it operates), the customers (including their local community, their social networks, friends and family, etc.), and the supply chain that helps the company create and distribute its goods and services (including the securing of raw materials, intermediary vendors, distributors, assembling organizations, service centers, etc.). You'll also note that the customer component has several areas that should be considered when you're exploring potential shared value; these are the parts of the consumption chain from the time at

Figure 5-1 Shared Value Framework

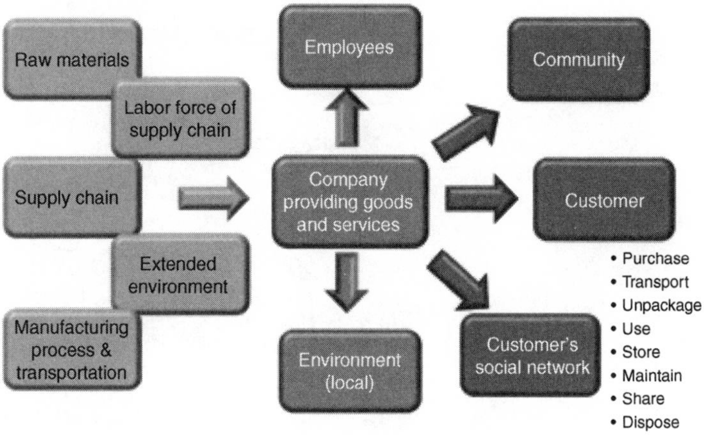

which the customer purchases the good or service, through transporting it, storing it, using it, sharing it, maintaining it, and eventually disposing of it. When the entire framework is considered, it is clear that there are myriad dimensions along which shared value can be created. This is where we can begin laying out the elements involved and seek opportunities and resources for shared value.

Keep in mind that we've been exploring some of the elements of this framework since many of these are members of the ecosystem. However, in this case, we are focusing on two members that underlie all ecosystems—society and the environment. Generally these two items are not taken into account when new products and services are being developed. This framework ensures that we formally look at ways of meeting the needs of these two entities as well as the standard ecosystem.

Let's go back to our Plumpy'nut example to map out a Shared Value Framework (shown in Figure 5-2). The organization that started Plumpy'nut (Nutriset), did an amazing job of creating social value. Not only were its goals focused on the direct needs of the target audience, but it considered virtually every other aspect of the framework. It employed

Figure 5-2 Plumpy'nut Shared Value Framework

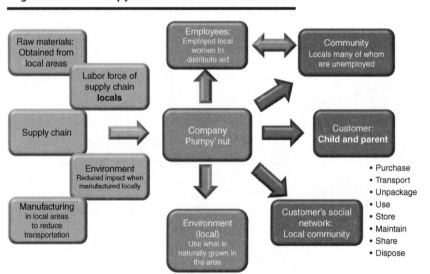

local women in communities as its method of distribution (creating new jobs for this community) and changed how malnutrition was measured, thus making it easier for illiterate people to identify which children needed help. Instead of using calculations from a height-to-body weight ratio, it simply used a midarm circumference measurement that could be easily taught to local people. It expanded its business by hiring local businessmen as franchisees for its operations, bringing the production closer to the need. This step further enhanced the social value generated by keeping more of the revenue in the local economy, creating jobs, using local raw materials, and reducing transportation. This organization exemplifies a company that is highly attuned to the society in which it is a part, and it has deep respect for the environment on which it depends.

To create shared value, a concept must provide value not only to the target audience but to society as a whole with respect to the environment and sustainability. Thus, as concepts are developed, they need to be evaluated with respect to these factors. The worksheet in Figure 5-3 provides a framework in which the needs of four key entities—customer,

Figure 5-3 Shared Value Worksheet

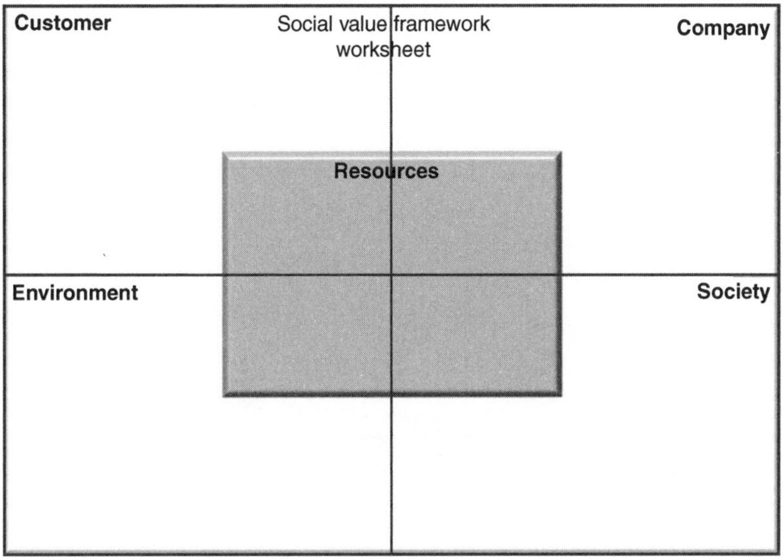

Customer	Social value framework worksheet	Company
	Resources	
Environment		Society

company, environment, and society—are mapped out with the resources that each entity can contribute. The resources include each entity's capabilities, actions, and processes over which they have control, to name a few. Identifying these capabilities and resources can be especially vital for overcoming key constraints that have not been resolved.

The worksheet has three primary uses: (1) concepts generated in the ideation sessions should be run through the framework to confirm that the concept has created social value and has small environmental impact (in cases where it does not, the worksheet provides a tool to stimulate thinking about features that might need to be added or adjusted to accommodate these needs); (2) existing products, services, and/or programs operated by the organization can be tested for their ability to create shared value, including where modifications may need to be made; and (3) the worksheet can be used to stimulate entirely new concepts that benefit the entire social framework, and, where possible, tap into resources that already exist.

Let's look at an example pertaining to resource conservation and sustainability for residential homes (see Figure 5-4). Two of the traditional ecosystem members are present—the customer and the company; however,

Figure 5-4 Shared Value Worksheet for Residential Resource Usage

Customer: Homeowner (buyer)	Social Value Framework Worksheet	Contractor or builder	Company:
• Ensure that home is energy efficient • Reduce waste in landfill • Reduce water used by household members	**Resources** Purchase power Control over usage of resources	Control over materials used Control over source of materials	• Sell homes that customers want • Maximize profit on the build • Avoid waste of materials
Environment: Construction site and surroundings	Raw materials Water Construction site land/materials	Regulations Incentives Peer pressure Outreach to masses	Society: Local community (existing residents) City and state governments
• Avoid depletion of nonrenewable resources • Avoid affecting the natural ecosystem, e.g., destruction of habitat, etc.		• Ensure that new homes do not create an environmental problem for the area • Control growth and sprawl	

society and the environment are added as a new part of the ecosystem. Let's look at some immediate solutions that come to mind from this initial mapping of the social value system. If we look at the capabilities of the contractor and the needs of the homeowner and society, it is clear that solutions that improve the ordering and management of supplies to reduce waste would be valued by homeowners, society, and the environment. Contractors can be incentivized to make their purchases from lumber yards that use sustainable harvesting practices through the influence of the local governments, peer pressure of other builders, and of course, homeowner demand. When considering the environment needs and the society resources or capabilities, cities could require new builds to have a plan that outlines the anticipated increase in water usage, where the water will come from, and whether this water source can sustain the additional development. Society and homeowners can also create demand for zero-waste home builds which will encourage contractors. Those that do build zero-waste homes can secure a differentiator in the market by targeting homeowners that value such an approach. Table 5-1 shows how the pairings created the ideas just listed.

**Table 5-1 Ecosystem-Wide
Solutions from Shared Value Worksheet**

Needs (who)	Resources (who)	Idea Generated
Reduce landfill (homeowner, society)	Process of construction, ordering of materials, control of use of materials (contractor)	Zero-waste certification for new home construction similar to the energy star rating for new homes. In areas where waste is important to homeowners, then this can become a point of differentiation for the contractors who can figure out how to do this. Local governments could offer incentives for these builds, especially if they save the city money in processing the waste.
Reduce wasted material from construction (contractor)	Incentives for low waste (society, local government)	

(continued)

Table 5-1 Ecosystem-Wide Solutions from Shared Value Worksheet (*continued*)

Needs (who)	Resources (who)	Idea Generated
Preserve nonrenewable resources (environment) **Create a point of differentiation (contractor)**	Control over purchase of supplies (contractor) Purchase criteria for new home (homeowner)— generate demand for low use of nonrenewable resources	Fair trade home— supplies obtained from lumber and other supplies from renewable resources or that execute replacement strategies— plant a tree for every one harvested, etc. This fair trade seal can become a point of differentiation for contractors that use these suppliers, much like Chipotle Mexican Grill promotes the suppliers of its meat, produce, and other products.
Preserve water (society, environment, customer)	Provide building permits (society, local governments) Evaluate feasibility of new building and development (society, local government)	Water-control development guidelines. Cities create guidelines for building permits that consider where the water will come from. They can also consider the current vacancy rates as well to prevent overbuilding.

As shown, mapping the needs and the resources yielded interesting possibilities that would contribute to a holistic approach to sustainable building that would resonate well with all members and truly create shared value.

Developing the Social Business Model

Once the concepts are run through the Social Value Framework to balance out the needs of society and the environment, it is time to put together the business model for the concept, specifically identifying how revenue will be generated, development financed, and operational costs covered. In reviewing different types of business models, I found an excellent book on the topic, *Business Model Generation*, which has created a versatile visual framework for mapping out many different types of business models with examples.[4] Although the book provides a thorough and effective tool for mapping out the relevant components of the business model, it is missing one critical piece—social impact. I found the versatility and visual aspects of the model very appealing and thus modified the *canvas*, as the tool is called, to include societal needs and the environmental impact of the concept, as well as the relationship and interplay between the members of the ecosystem (see Figure 5-5). The restructuring of the canvas also helps tie the categories to the work already completed in our Shared Value Framework. This simple canvas can be used to clarify and work through the details of a business model for a new concept. Please note that if your organization already has a business model framework, feel free to use whatever you are comfortable with. However, since most business models do not contain society as a customer, nor do they include the environmental impact, these categories will need to be added.

A business model canvas should be created for each of the concepts included in the road map generated by the Ideation Portfolio. The customer

Figure 5-5 Social Business Model Canvas

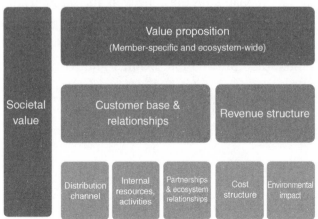

base and relationships for the business model are taken straight from the Ideation Strategy Guide as well as the Ideation Portfolio. Thus, in the Plumpy'nut case, the customer base is "parents in rural areas that have high levels of malnourished children" and the "clinics that purchase the food aid." This section would also include the important secondary relationships including the clinic workers, relief workers, and local friends/family members of the malnourished child. All of this information is found in the Social Innovation Blueprint.

The societal value is a culmination of the needs and constraints of the entire ecosystem that have been addressed by the new concept. The social value can also include projected improvements in the social issue and projected ROI; for Plumpy'nut, how many children are expected to be saved, how much money will be saved compared with treating them with an IV in a hospital setting, the efficacy of the concept, and so on.

The value proposition should be created for each of the target audiences; that is, there will be a value proposition for the parents and another for the aid agencies that purchase the food. It is important to incorporate a number of different elements including the features of the concept that meet the Hi-Op Needs and/or the Hi-Impact Constraints, the emotional jobs that the concept satisfies as well as any human constraints that it overcomes. All this contributes to a true data-driven value proposition.

The bottom row of items in Figure 5-5 are the internal components— what the organizations participating in the delivery of the solution will need to provide, what resources must be committed, what activities must be executed, and how they will be coordinated among the sectors for implementation. The distribution channel and partnerships are something to be discussed and debated to identify the optimal approach. In the Plumpy'nut case, the distribution channel consisted of the local people who became the distributors, suppliers, and executors of the program.

The cost structure in this model also includes the environmental impact. When Plumpy'nut began, it had a centralized manufacturing facility that was cost-effective from an operational efficiency perspective; however, when ancillary environmental costs were added into the equation such as transportation of the raw materials and the final product, it was not as low cost as originally thought. The organization shifted to a decentralized, local production model and saw immediate financial savings and reduced environmental impact (reduced transportation). Not to be overlooked was the economic impact on the community in which the product was used; jobs were created, local raw materials purchased, and the money stayed in the community.

The revenue structure is usually tough to work out. Who will pay for the new concept? How will its development be funded? We review approaches to funding in the next section.

Overall, this framework's most significant value is that it forces the innovation team to think through the value of the concept to society as a whole and its impact on the environment before ever launching a new program, product, or service. This simple practice, if propagated through every sector—business, government, and nonprofit—would generate great value for our world.

Business Model Ideation

In many cases, the concepts that have been created at the ideation session still have one more level of creativity to go through—the revenue and cost part of the business model. You have a great concept and know that it will address important needs for the key members of the ecosystem. The question now is, how do the dollars flow?

As we know, there are three areas in which to create an innovative business model—revenue, fixed cost, and variable cost. A set of business model triggers has been designed by Ellen Domb and Tony Ulwick, based on the TRIZ methodology and other creativity techniques, that provide excellent stimulation for looking for business model solutions. Some examples of each trigger type follow.

Business Model Triggers—Variable Cost

The variable cost of a concept is one of the first places to be creative by finding unique ways to get the job done but to pay much less, if anything, for it. This requires identifying what resources are available within the ecosystem that might have value for another member of the ecosystem. Many nonprofits make excellent use of volunteers, which is a great way of eliminating a variable cost. A sample of variable cost triggers is listed below:

1. Pass a cost entirely on to the customer; for example, have customers sell, assemble the product, install the product, maintain the product, and so on.
2. Obtain supplies and materials for free; for example, use someone else's waste as a raw material.
3. Make labor free; for example, consider the use of volunteers, interns, nonprofits, academics, and so on.
4. Reduce costs by using existing infrastructure for multiple purposes; for example, other products, markets, operations, and the like.

As an example of trigger 2, in many parts of the country, dairy farmers can now earn revenue from the waste generated by their dairy cows. Instead of the waste becoming a cost (to scoop up and haul off the waste), it is now sold to electric energy producers who use it as a raw material. This provides a variable cost reduction to the energy producers and a new source of revenue for the dairy farmers, hitting two different categories of triggers.

Business Model Triggers—Fixed Costs

There are many ways to be creative with fixed costs. For example, in Austin, Texas, a company called Car2Go saves people the expense of buying cars and eliminates the hassle of renting cars. It is designed for people who live in an urban area and rarely need a car. However, when they do need a car, they can simply pick one up at a Car2Go location. This type of activity, where expensive items are being leased in order to share them with others, is becoming more common with boats and even houses. Other examples of fixed cost triggers are listed below:

1. Use existing resources in a way that eliminates capital investments; for example, facilities, equipment, tooling, and so on.
2. Reduce fixed costs by incurring them when they are the lowest; for example, when subsidized, when financial incentives are in place, and so forth.
3. Get someone else to pay for capital investments.
4. Leverage a capital investment made by somebody else; for example, an intermediary, a third party, and so on.

Business Model Triggers—Revenue

One of the more challenging, but interesting, innovations takes place with the revenue itself. Many companies have applied these techniques to make major breakthroughs. In the commercial area, Netflix is a prime example of using a new business model—videos direct to the consumer via mail for a monthly fee. This changed the way people rented videos in the United States. Many of its competitors have gone out of business. Some of the triggers for the revenue side of the equation are listed below:

1. Take a revenue source from someone else in the value chain; for example, supplier, distributor, installer, maintainer, and so on.
2. Identify resources inherent to the value delivery platform and sell/use them; for example, information, energy, and the like.
3. Collect revenue that forces the generation of future revenue.
4. Accept a noncash form of payment; for example, credit cards, a subscription, royalties, and so forth.

5. Collect revenue from a different payer; for example, insurer, distributor, advertiser, government, and so on.
6. Collect revenue at a different point in time or different frequency.
7. Generate revenue from operational activities; for example, sell tours to witness operations, have outsiders buy your company training, and so forth.

A fun example of generating revenue from operational activities is a group called Vocation Vacations. Its service allows individuals or organizations to sell on-the-job vacations to others interested in its industry. Winemakers, chefs, artists—professionals in dozens of industries—generate revenue by allowing others to work for them for a couple of days. This provides revenue for the host since the vacationer pays for the experience of watching (and in some cases, helping) the host conduct their day-to-day business.

New Financing Methods for Social Innovation

Collective impact is the "commitment of a group of important actors from different sectors to a common agenda for solving a specific social problem."[5] This movement of bringing together everyone involved in the social scenario in order to drive significant change is demonstrating good results in the education sector. The challenge seems to be in getting funding. Because these initiatives are not focused on the development of a single product or program, they are often difficult to get funding for. There needs to be a fundamental shift in how funders see their roles—they must begin to see that they are funding an entire scenario, a complex set of jobs that may require several solutions, products, and components in order for success to be achieved.

In an article called "Catalytic Philanthropy," the authors suggest that, "The most powerful role for funders to play in addressing adaptive problems is to focus attention on the issue and help create a process that mobilizes the organizations involved to find a solution themselves."[6]

Job-Based Funding

In many cases, organizations seek grants and philanthropic funds for start-up costs. The value of the data from the social innovation program that created the concept cannot be underestimated. This information will provide a solid business case for pursuing grants and other contributions to get the program off the ground. The data will provide exact details on how the concept is expected to meet the needs of the people and will show how the concept will overcome the environmental, political, and cultural constraints.

This is especially important with the growth of collective impact models where numerous entities are all working together to address a complex wicked problem. These entities are bound together to achieve an end result; however, in the initial phases the solution to be funded is not yet conceived. If they are truly conducting a needs-first approach, then the first step is to collect the information, as we've pointed out in the earlier chapters. Once the needs are established, it is the solution to those needs that is being funded.

Thus investors are not supporting a specific product, service, or program— yet. Instead they are investing in the *job to be done*, the unmet need that will be resolved by the innovation team.

Pay for Success

In 2009, President Obama requested $50 million to identify and expand effective nonprofits. First Lady Michelle Obama stated in a press release:

The idea is simple: to find the most effective programs out there and then provide the capital needed to replicate their success in communities around the country that are facing similar challenges. By focusing on high-impact, result-oriented non-profits, we will ensure that government dollars are spent in a way that is effective, accountable and worthy of the public trust.[7]

In this pay-for-success model, social programs would need to first obtain money from other sources—foundations, donors, private investors, and so on. Then once the success of their method is proven, the organization behind the social programs would receive government support. This approach incentivizes successful programs. If the programs do very well, the original donors may get more than their original investment back.

The state of Massachusetts recently started a social innovation financing program that would allow it to engage in innovation programs and performance improvement of the government programs that are already in place. "Social innovation financing is a creative idea based on a simple premise—have government pay for demonstrated success rather than the promise of success," said Secretary of Administration and Finance Jay Gonzalez. "It's an approach concept that we hope will result in new and better ways of providing critical state services, achieving better outcomes, and stretching every taxpayer dollar as far as possible."[8] If this program continues as planned, then Massachusetts will be the first state to implement a full social innovation financing program with social impact bonds and pay-for-success contracts.[9]

Obviously, a key aspect of the pay-for-success program will be the performance metrics on which it is judged. This is yet another reinforcement of having the needs data up front, supporting the idea and the program.

Social Impact Bonds

Social impact bonds (SIBs) are another new financing method that support better outcomes in social initiatives. The SIB is an actual contract with the public sector to pay for improved social outcomes that result in public sector savings. The anticipated public sector savings are the basis for which the funds are raised, the fund itself is focused on prevention and early intervention services that would yield the anticipated improvement in the outcomes. They are unlike conventional bonds in that they do not offer a fixed rate of return. Instead, the return is based on the performance and savings achieved. SIBs started in the United Kingdom, and much interest is being generated in the United States. An organization in Boston is offering these bonds in the United States, and in President Obama's 2012 budget there is a proposed $100 million to be used to run SIBs.

The benefit of these types of bonds is the availability of cash to focus on prevention and early intervention programs. The impetus is on the investors to ensure that the programs will really produce the results that are projected. As such, it would be wise for a methodology such as the Social Impact Framework to be used to articulate the actual problem, the potential for improving the outcomes of very specific jobs, and the ability to measure the success effectively before making such an investment. We recommend that the methodology be implemented as due diligence prior to the funding of a SIB.

Water Fund Trusts

A recent program launched by the Nature Conservancy, a nonprofit dedicated to protecting nature and preserving life, is a Conservation Water Trust Fund originally launched in Colombia. The Water Trust Fund is an innovative financial investment tool that "receives contributions from Bogotá's water treatment facilities to subsidize conservation projects—from strengthening protected areas to creating incentives for ecologically sustainable cattle ranching—that will keep sedimentation and runoff out of the region's rivers." The funds are then distributed in programs that will ensure preservation of the purity of the water in the area including helping locals learn about forest conservation, financing changes to ranching practices, and helping farmers update their equipment and sometimes even modify their crops to drastically lower water usage. The trust fund was established by the conservancy to protect the watersheds and rivers in Colombia "from strengthening protected areas to creating incentives for ecologically sustainable cattle ranching."[10] This fund is working actively with local farmers, businesses, scientists, conservationists, community members, and many others in an attempt to produce a sustainable environment to prevent what could be a disaster resulting from water depletion.

The extent to which the conservancy has involved all the major ecosystem members is key to its success. It has involved the locals and the owners of large businesses. And it has directors to coordinate the program. The large water treatment facilities subsidize the conservation efforts to ensure that the pollutants never reach the rivers and watershed. Instead of spending millions to remove the pollutants, the investment in conservation is made up front. "The Conservancy brought together a broad range of public and private stakeholders—many of which had never collaborated before—to . . . broker the landmark agreement."[11] It is projected that the savings to the water treatment facilities will exceed $4 million per year. (Please note that this information is also contained in Chapter 8: Conservation.)

This complex ecosystem requires significant coordination and balancing of needs and has the potential for conflicting needs. To address this, the conservancy has local representatives who educate the local merchants, farmers, and ranchers on the need for the changes, the impact the changes will have, and the dangers of not implementing the changes to the management of the natural resources of the area. According to Aurelio Ramos, director of the Conservancy's Northern Tropical Andes Program, "In financial terms and in actual conservation results, these funds yield a tremendous return on investment."[12]

In Summary

Social innovation is a movement in its infancy and as such struggles to find financial models that work. The increasing visibility and attention of social innovation at the highest levels of governments around the world is an encouraging step forward. Once governments, nonprofits, and the corporate world begin to focus on solving problems together, realizing that there truly is shared value to be had by addressing problems, the faster social innovation programs will take root and prosper. As stated by Porter and Kramer, "Shared value holds the key to unlocking the next wave of business innovation and growth. It will also reconnect company success and community success in ways that have been lost in an age of narrow management approaches, short term thinking and deepening divides among society's institutions."[13] In short, adopting a philosophy of shared value can not only stimulate our lethargic economy but can also produce long-term social benefits as well.

Part | 3

IMPLEMENT THE SOLUTION

Chapter | 6

DIFFUSION OF INNOVATION

A s we get to the final step of the methodology, we face one of the most difficult challenges yet—the diffusion of the solution. Susan Evans and Peter Clark argue that, "Spectacularly effective social innovation programs often fail to take root in other places. The social sector . . . seems to have less enthusiasm for mastering the skills of transplanting successful innovations to other needy locales."[1] Diffusion of innovation is generally a difficult process, one that is fraught with demanding issues. Studies of successful diffusions of innovations often require several key features:

- The innovation directly addresses an unmet need of the intended target.
- Low capital investment per unit of output and per employee.
- Organizational simplicity.
- High adaptability to a peculiar social and cultural environment.
- Use of local natural resources.
- Low cost of final product.
- High potential for employment.[2]

The ability of an innovation to meet these needs greatly increases their chances of a successful diffusion into society.

Orphan Innovations

The failure of a fantastic innovation for feeding the poor became the catalyst for understanding how to effectively diffuse innovations into new regions. Mickey Weiss owned a produce wholesaling business and saw that there was tremendous waste composed of produce that did not make it from wholesale to retail within its viable shelf life. Unbelievably, this produce was destroyed; literally tons of food was going to waste because it was past the date it could be shipped to and shelved in stores, and yet there were thousands of homeless and hungry people within the local community. Mr. Weiss's innovation was to get this food to the local food banks which could then offer fresh healthy produce, in addition to the staples and canned items, to the hungry they serviced.

Mr. Weiss created a process for procuring the food, not only from his business but from other wholesalers, and he distributed it to the local Los Angeles food banks. The program was very successful in his area and received national attention as well as great local press. It would seem that its diffusion to other parts of the country would be a given. Mr. Weiss and researchers Evans and Clark found that it was more difficult than just having a great idea. In fact, they coined the term "orphan innovations" to refer to the large number of innovations that simply never get beyond initial implementation.[3]

The key problem with most attempts to diffuse an innovation is a lack of understanding of the needs and constraints of the *new* market—they are assumed to be the same as the initial introduction market. The new food bank program developed by Weiss was assumed to be desired by other area food banks, but the constraints and member needs of other markets were not considered and were found to not be consistent with the situation in Los Angeles. "We were attempting to transplant a *copy* of the Los Angeles program. Instead, we should have been transplanting the *concept of produce collection* in whatever manner the local food bank could best use. We had underestimated the importance of embracing variation in each site's geography, business culture, charitable infrastructure and more (emphasis added)."[4] This example provides excellent support for the job-based application to social issues: needs must come first, with the solution derived from and modified with an understanding of the needs, even if it is only a shift in the geography or demographic. In this case, collecting produce and distributing it to the hungry are the key jobs that should have been studied in the new markets before implementation was attempted.

As Weiss, Evans, and Clark went through the process of trying to get other cities interested in executing a similar program, they found numerous constraints, both human and environmental, that affected the adoption of the program. The biggest constraints involved the mindset of the targeted extension food banks:

1. Food banks often felt "embattled and defensive, believing that the general public was complacent about poverty" and did not believe they would be able to get the support they would need to obtain and distribute the fresh food.

2. The food banks lacked a general understanding of how the wholesale/retail produce business operated and thus did not believe that wholesale firms would hand over food that was still good for consumption.

3. Food banks often believed that the beneficiaries of their program, impoverished people, would not be interested in fresh foods given the availability and familiarity of convenience foods that they traditionally received from the food bank.[5]

With this understanding of the constraints, the Weiss, Evans, and Clark team was able to put together a program that overcame these issues, educating the new food banks in the process and helping them understand how to reposition the food to the population they served.

The authors provide a framework that suggests the optimal method of disseminating new solutions is through careful customization of the solution. Instead of simply trying to replicate the solution in another locale, it is important to understand the factors that make the new locales unique, provide a plan to overcome unique obstacles, and then tailor the solution in a way that best meets the needs of the new locale. This approach is very similar to the plan outlined in the remainder of this chapter. In our vernacular, proper diffusion requires that the needs and constraints of the new ecosystem members be quantified and prioritized. Then the solution, and possibly the business model, should be adjusted accordingly.

Social Impact Framework's Diffusion of Innovation

The process illustrated by the Social Impact Framework can be applied in an abbreviated form in order to maximize the success of taking a solution to a new location or new population. The situation needs to be framed to help people understand the nuances of the problem. The needs and constraints of the population must be understood and prioritized. Their Hi-Op Needs and Hi-Impact Constraints must be examined to enable understanding of the underlying issues. And the concept needs to go through some further assessment with respect to these needs; it must also go through possible ideation to revise components. Finally, the business model must be evaluated to aid in understanding whether the same model will work in the new location. The business model is an especially important component, because the new location may require changes in distribution channels, suppliers, partnerships, and the like.

Step 1: Define the Solution

The first step involves going back to the framing guide and the Social Innovation Blueprint to systematically validate the situation in the new market, identifying any changes in executor, beneficiary, potential new constraints, third parties, and so on. This may be especially important if the innovation is being taken to a new country where there may be new regulators involved, new intermediaries, different support agencies, and the like. Are the focal jobs the same as those on the original blueprint? Are the contexts the same, or are there changes? Are there new constraints—environmental or human—that need to be considered? Again, when crossing locales, it is vital to determine whether there are cultural or social issues that could impede the adoption of the innovation. What about the existing platform— are there significant changes in the new locale? If so, how could those changes affect the successful implementation of the innovation?

Step 2: Identify and Prioritize Needs

The list of needs obtained in the original initiative must be reviewed carefully to identify any potential problems and to validate the strength of the opportunity in the new area. Note that this does not have to be as large an investigation as the original; it can be handled with a small number of interviews of various members of the ecosystem to validate the job steps and constraints, and a smaller, but readable, sample (100–300 people, depending on your comfort level with margin of error and budget) for the quantitative. Keep in mind that the time and money expended in this exercise will be directly offset by the cost of a failed implementation in the new market. If bringing the innovation to the new market will take a sizable investment of time and money, then the time spent executing these steps pays excellent dividends in confirmation of the plan and in peace of mind.

The following actions should be executed.

1. *Validate the focal job.* Do the members execute the focal jobs differently in the new location? This could affect the job steps and may require a few additional qualitative interviews in the new location to validate the job map or modify it as needed. In fact, the existing job map and job steps can be shown directly to the ecosystem members in the new location, with a request for them to validate it or indicate any differences.

2. *Validate the wording of the need statements.* Are the need statements written in a way that is consistent with the new location's culture? Sometimes the need statements may refer to an object or activity that doesn't exist in the new location. For instance, a need statement may refer to a government regulation that exists in the United States but that wouldn't apply elsewhere. There could be differences in common terminology, and in some cases translation is necessary. It is important that the needs are reviewed for localization by someone in or very familiar with the area.

3. *Verify the constraints.* It will be important to interview several people from the new area, in various parts of the ecosystem (job executors, overseers, beneficiaries) to validate the constraints from the original study and to identify any new ones. This is a very important step since the assumption that the constraints are the same in the new area can have a significant impact on the success of the implementation.

4. *Prioritize the needs in the new area.* Importance and satisfaction are likely to differ depending on the solutions that exist in the new area. As such, the prioritization of the needs must be reestablished for each new area being considered for the innovation. This step is critical for three reasons. First, you will be able to verify that the needs are strong in the new area and that the constraints can be managed. This should be taken care of before you spend time and effort to bring the innovation to the new area. Second, this process can be used to prioritize among several different areas being considered for the solution by identifying those areas that have the greatest opportunity based on a small but readable sample in each of the areas you are considering. Third, the data will identify any modifications to the solution that need to be made in order for the concept to be attractive in the new area. Alternatively, it is also possible that there may be items in the solution that are not needed in the new area which may allow a less complex and costly version of the solution to be diffused to that area.

Step 3: Examine the Opportunity

Once the data have been requantified in the new location, identify whether there are any major differences in the opportunities. Are any changes required in the concept? If so, in what way? For instance, is there a new need that comes up as a Hi-Op Need? Did one of the original Hi-Op Needs *not* rate high in the new area? This could pose a problem for adoption if the solution hinges on this particular need.

Constraints must also be examined for differences in impact and control which are very likely to exist in the new market. These changes in the constraints should be reviewed to ensure that the solution can be implemented in the new market, or to identify necessary modifications to the solution, the business model, or the implementation plan in order to accommodate the differences in constraint ratings.

It is also imperative to reevaluate the context—the reasons behind the high opportunities—to identify anything new or unusual. Again, this could be just a handful of interviews with different members of the ecosystem. The context differences will be important for identifying any changes needed to the messaging or rollout in the new area.

Step 4: Review the Solution and Modify It If Necessary

In many cases, the solution created for the original innovation will require some tweaking for the new area. To maximize the adoption of the innovation, customizing it for the new locale, to better meet the local needs, will be vital. For example, if the food bank program explored a new city for expansion and found that the population the local food bank serves really doesn't have an interest in healthier foods, then the food bank program may want to reconsider that location. Or the solution may require an additional component such as an educational outreach to get the population to understand the value of eating fresh food every day.

The key is to ensure success in the diffusion of the innovation. Just like the original innovation, this will take some time and effort to complete.

However, some of these programs only have one shot to be successful, and, therefore, it pays to make that one shot hit the target directly.

Step 5: Review the Business Model

For the diffusion to be effective, we need to understand whether there are any significant changes to the business model. If the food bank innovation we discussed was brought to a city that had a waste disposal fee for dumping produce that wouldn't make it to the retail market within its shelf life, then this market will have an added benefit for the wholesaler on the cost side of the equation and can become further incentive for the wholesaler to participate. There may be other key partnerships or other key activities that need to be considered in the new area. There may be other channel options, and there may be significant differences in the customer relationships between the food bank and the recipients.

The customized business model for the new area can be an excellent tool for bringing new organizations onboard. In the food bank example, the benefits outlined in the business model are straightforward and easy to see. If this business model is also backed up by the needs analysis for the market, it is much more likely to get serious attention from the wholesalers in the new area. The business model is a great selling tool for expansion of innovations into new areas, seeking funding for the expansion, or providing feedback to investors as to expansion potential.

Scaling Impact

In a recent *Stanford Social Innovation Review* article by Jeffrey Bradach, the author discusses his theory on the impact of scaling nonprofits. He notes that while many organizations dismissed the idea of scaling their solutions as impossible and "too corporate," there were many others who were building solid replicable models, expanding their efforts geographically, and becoming some of the largest nationally recognized nonprofits such as Teaching for America and Habitat for Humanity. "These organizations have found that scaling is anything but an exercise in cutting cookies, as it requires not only fidelity to core processes and programs, but also constant adjustments to local needs and resources."[6] As Bradach points out, the goal now is for organizations to determine how to get a 100 percent increase in impact with only a 2 percent increase in organizational size.

Some of the expansion strategies identified in the article include the utilization of the Web, formal or informal networks of organizations, and the use of intermediaries. The extensive use of Web tools can make "programs in a box" available with the instructions, tool kits, and guidelines for other locations to borrow the basics of the idea and run with it in their community. This can include not only a strong website with associated tools, but the use of social media to spread the word about programs, attract local volunteers, and get grassroots movements underway. Large networks can be established with a strong centralized organization (such as Boys and Girls Clubs of America) or be loosely organized around a philosophy and model such as Alcoholics Anonymous or the hospice movement.

Intermediaries are another good scaling option that can provide knowledge, tools, best practices, and performance benchmarks to other organizations. NITLE (National Institute for Technology in Liberal Education), for example, provides a collaboration—an idea and knowledge sharing partnership to small to mid-sized liberal arts college programs that don't have the resources of their larger counterparts. Through NITLE liberal arts campuses can, "Develop local and collaborative approaches to integrating technology into teaching and learning, managing infrastructure and planning for the future."[7]

The tools outlined in this book can assist expansion efforts by providing the new locations with significant information on the ecosystem, a tool to immediately begin localizing and prioritizing the needs in the new environment, and an understanding of the needs that the solution helps to address. This framework should assist in launching innovation programs in new locales following a systematic, strategic method, allowing customization to take place in a way that will directly target the needs and constraints of the new location.

Additionally, organizations should seek to establish a collaboration dataset that allows the disparate locations to document ideas that helped address the needs in their area. That way as new locations come on and identify key areas of need, they can review the database and find out what kind of solutions have been implemented for these or similar needs in other locations. Such additional value will help the new potential organizations start with even more effectiveness.

Maximizing the impact is not only accomplished through geographic and physical expansion, but by changing perceptions, attitudes, and behaviors. We call this, "Moving the importance needle." In some cases, the issues that organizations struggle to break through are not on the radar of the individuals who can initiate and activate change. Fortunately the social impact model provides immediate insight into the presence of this problem via low importance scores. If an organization is working on preventing childhood obesity and executes a survey with pediatricians and parents and teachers of elementary school children and finds that key aspects of the job of preventing childhood obesity are *not* rated important by the parents, then we know that we have an educational issue or a denial issue. In these cases, the focus will have to be placed on changing the perceptions and increasing awareness of the problem.

Another example of this phenomenon was prevalent in a recent study we conducted on health care in a small service area within the eastern United States. We found an astounding lack of high importance rating on any job pertaining to managing health and well-being. Not one job was rated as a high opportunity for these people. Not managing their diet. Not determining how much exercise was needed. Nothing. We had executed this study in several other geographic areas, and these job statements usually were rated very high. This group, however, did not find these jobs to be important. But they did find the jobs around obtaining health care when they were sick very important. The prevailing attitude was basically one of, "Fix me when I break, but don't ask me to take preventative measures." As you can imagine, if you are a health care system that is trying to manage costs, this is not a good attitude. The organizations in this area will need to spend time and money educating this population about the necessity of maintaining their health to avoid the increasing costs of fixing the problem.

As Bradach points out in his article, sometimes it takes a successful innovation *outside* the system to drive change inside the system. He cites the example of charter schools that, through their success, have shown that change and improvement in education really is possible. "By demonstrating that all kids can perform well if given a good education, ... charter organizations have transformed the debate about what people can and should hold schools accountable for."[8] This is a critical strategy for disruptive innovations and is discussed more in Chapter 9.

In Summary

Orphan innovations are unfortunately the norm rather than the exception. The key dynamic creating these orphans is a faulty assumption—the assumption that an innovation that works great in one place will work great somewhere else. While the core of the innovation is likely to be valued in a new location, it comes down to the unmet needs of the new location and the constraints that exist in that location. By applying the first five steps of the Social Impact Framework, the success rate of innovation diffusion will dramatically increase since the solution will be tailored to meet the needs of and overcome the constraints of the new location. Additionally, by reviewing the business model components and validating these in the new location, the financial risk of the diffusion is also alleviated. There is nothing sadder than a great idea that does not meet its full potential. It is the responsibility of today's social innovators to ensure that the orphan idea epidemic becomes a historic phenomenon, no longer plaguing the field of innovation.

Chapter | 7

HEALTH CARE

The health care system contains some of our country's most significant technological achievements, and yet millions of people go without adequate health care, often because they cannot afford to be insured or to pay out of pocket for the services. Attempts to solve the health care crisis in the United States have met with strong resistance—especially when solutions are mandated that violate the needs of other members of the ecosystem—and continued failure. Nearly every president over the past two decades has attempted legislation that would create major overhauls to the system so that health care would be more affordable. None of these programs have succeeded as of yet.

There are two primary issues that plague the health care system, making it difficult to be optimized:

1. *A single high-level platform.* The health care system is typically addressed in its entirety as if it were a single entity instead of the separate jobs it is composed of. The health care system was designed to address the most complex medical conditions with advanced technology, highly trained professionals, and an infrastructure that included

general hospital systems, individual medical professionals or practices of multiple physicians, and state-of-the-art diagnostic and treatment equipment. The platform is perfectly tuned to address a specific type of job—those that fall in the category of complex medical problems; however, as such, it is also overkill for a vast majority of less complex jobs (other health care issues) that are more rules-based, easily diagnosed conditions. The platform is both insufficient and overkill for wellness and prevention; insufficient by not having the programs and reimbursement mechanisms for wellness, and overkill because the platform was designed to *fix* large-scale problems, not prevent them. Wellness jobs have been forced to fit onto the platform and are not well served by them.

2. A *single business model.* Different jobs, with different levels of complexity and different definitions of quality, *require* a different business model. However, the U.S. health care system applies the same business model to all of the vastly different jobs, again creating a situation of overserving a large percentage of the jobs. "The lack of business model innovation in the health care industry—in many cases because regulators have not permitted it—is the reason health care is unaffordable."[1] All in all we have the most advanced health care system in the world and it is bankrupting our country.

Define the Problem

As shown in Figure 7-1, the health care system is an incredibly complex ecosystem crossing all sectors—corporations, nonprofits, and government entities—and is based on six major groupings: the hospital system, patients, employers, physicians, payers, and the government regulatory agencies.

From the complexity of the ecosystem itself, it is easy to see why this social issue has evaded resolution for such a long time. When the jobs are added to the equation, as well as related jobs, emotional jobs, and constraints, it is easy to see how this problem has become so

Figure 7-1 Ecosystem of Health Care Scenario

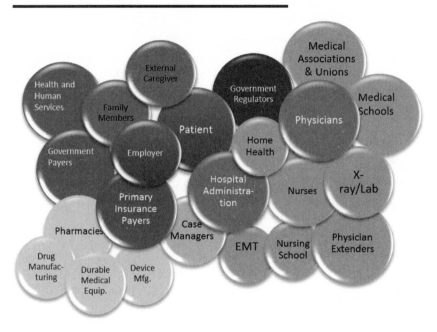

intractable. Trying to balance the job satisfaction in an ecosystem of this size is a huge undertaking and should be addressed job by job.

Blueprint

As discussed earlier in this book, the key to solving wicked problems is to clarify the problem definition. The Social Innovation Blueprint can be used to parse the jobs and contexts for specific health care scenarios. This type of problem definition is the first critical step to creating new value within the system as a whole.

In the Social Innovation Blueprint in Figure 7-2, there are three contexts—maintaining health and well-being, obtaining treatment for an illness or injury, and recovering from the illness or injury. Once we begin to lay these into the blueprint, the jobs, roles, and interactions become clear.

Each of these contexts defines a shift in the type of job being accomplished and thus is likely to include new members of the ecosystem and

Figure 7-2 Social Innovation Blueprint for health care

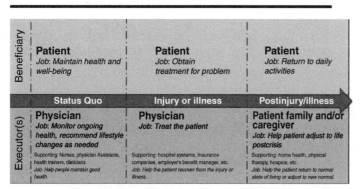

new jobs that the members perform, as well as a shift in platform and business model. For example, during the status quo, the focus of the patients and their encounters with the health care system is wellness and prevention. Once an acute or chronic health care situation takes place, the focus shifts to care for the problem and away from health/wellness. Then there is the postincident period where there are even more players added, depending on the severity of the condition, and new jobs to be undertaken. Thus, at a minimum, these three contexts must be explored in addition to the handoffs that occur during the transition between these contexts.

The injury/illness context is generally specified by the job category of the injury/illness. For instance, a new blueprint should be developed for health care incidents that involve acute/short-term episodic conditions, chronic health conditions, or terminal/critical illness or injury. These would involve different jobs and potentially different or new job beneficiaries or executors.

For example, within a single job category such as chronic diseases, there are two phases—the acute phase of the condition and the nonacute phase, which is the adherence phase designed to maintain the condition. In the past, many of today's chronic conditions were acute, often fatal, and the systems and business models were set up to treat them much as cancer or other terminal illnesses are today. However, because of effective

Figure 7-3 Multiple Context Blueprint

	Beneficiary Job	Beneficiary Job	Beneficiary Job	Beneficiary Job	Beneficiary Job
	Monitor for stroke signs/symptoms	Initial stroke symptoms—transport	Emergency treatment	ICU Critical care unit	Long–term care (home or nursing facility)
Executor(s)	Executor Job / Supporting executor(s) Job	Executor Job / Supporting executor(s) Job	Executor Job / Supporting executor(s) Job	Executor Job / Supporting executor(s) Job	Executor Job / Supporting executor(s) Job

treatments, these chronic conditions can be treated and put into disease management—but they must be actively managed.[2] In this example, the job beneficiary becomes the job executor in the phase of adhering to the prescribed treatment plan.

Let's look at an example of a critical illness scenario—a stroke patient—being viewed by a hospital system that is creating an innovative Stroke Care Center. In this example, there are many contexts that must be understood in order to manage the entire condition from onset of symptoms through rehabilitation. The blueprint in this case is created from the status quo (prestroke or patients with a predisposition to the condition), the treatment of a stroke episode from the ER perspective, the ICU perspective, the immediate recovery perspective, and long-term care perspective. This would yield a multicontext blueprint such as the one shown in Figure 7-3.

Identify Unmet Needs

Once the blueprint is mapped out, the next step is to identify the unmet needs of the major job beneficiaries and executors, as well as the environmental and human constraints. In the stroke center example, there are several job maps to be investigated; Table 7-1 illustrates the focal jobs

Table 7-1   Focal Jobs for Stroke Center Ecosystem Members

Member	Beneficiary	Executor 1	Executor 2 . . . n
Monitor for stroke signs/symptoms	Patient Job: *Prevent a stroke from taking place* *Determine if stroke symptoms are occurring*	Patient/patient's family Job: *Determine whether the patient is experiencing a stroke*	Primary care physician Job: *Monitor patient for factors that could lead to stroke; educate patient and family members on stroke symptoms*
Initial stroke assessment— transportation	Patient Job: *Seek medical treatment*	EMT Job: *Assess the patient, stabilize the patient, transport the patient to the hospital*	Emergency room physicians Job: *Receive the patient, conduct initial treatment of the stroke*
Emergency treatment	Patient and family Job: *Obtain emergency treatment* Family: *Provide information to the physicians on symptoms, onset, etc.*	ER physicians, neurologist Job: *Determine the type and severity of the stroke; determine best course of action; stabilize patient*	Nurses, lab and CAT scan staff, etc. Job: *Execute necessary tests to determine stroke type and severity; coordinate care with physicians*
ICU/critical care unit	Patient Job: *Submit to treatment* Family: *Authorize treatment, check on patient's status*	Neurologists Job: *Treat the stroke, communicate with patient's family*	ICU nurse, stroke specialists Job: *Manage the stroke patient's care*
Long-term care	Patient Job: *Adhere to treatment plan* Caregiver Job: *Support patient in executing treatment plan*	Neurologist Job: *Determine the recovery potential of the patient; adjust treatment as needed; communicate with primary care physician*	Primary care physician Job: *Monitor patient's progress; assess impact on other comorbidities* Home health nurse, physical therapists, etc. Job: *Help patient recover functionality*

in each of the contexts, these must then be analyzed for the detailed job steps. It is a very complex job.

As mentioned, each of these focal jobs will need to be investigated further along with the related jobs, constraints, emotional jobs, and so on. Table 7-2 breaks out the single box from Table 7-1 "ICU nurse: manage the stroke patient's care" and details some of the job steps, related jobs, and so forth.

Once the needs are gathered from the various members of the eco-system, the prioritization of importance and satisfaction of each of the need statements is uncovered. To illustrate the opportunity spectrum for a health care project, let's review the information on uninsured workers in the state of California (Figure 7-4).

The study we conducted in 2006 for the California Health Care Foundation yielded a number of interesting findings. At the time of this study, the retail health care models were just getting off the ground. When we look at the "episodic" conditions that these retail models are focused on, we can see that within the uninsured population of California the medical conditions that the retail models address are ready for a solution—their opportunities are well over 30 percent. Some of the high opportunities

Figure 7-4 Opportunity Spectrum for Uninsured Workers in California

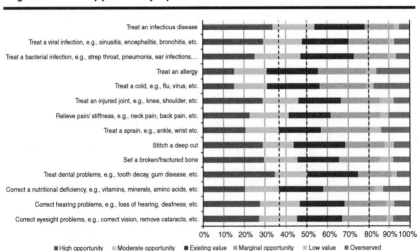

Table 7-2 Treat a Stroke Patient

Focal Job: Manage the Stroke Patient's Care
Member of Ecosystem: ICU Nurse

Job Map Step Name	Job Step
Define	• Evaluate the patient's condition
	• Review the physician's orders
	• Clarify any confusion in the orders
Locate/prepare	• Order prescribed medications
	• Order prescribed diagnostic tests
	• Prepare the patient for the diagnostic tests
Confirm	• Ensure that correct medication is given to the patient
	• Ensure that the correct tests are conducted on the patient
	• Ensure that the orders are complete
Execute	• Execute the treatment plan
	• Administer medications
	• Coordinate information between shifts
	• Coordinate care between doctors
Monitor	• Determine how the patient is responding to treatment
	• Notify the physician if there are changes in the patient's condition
Modify/conclude	• Conduct any changes in the treatment plan
	• Document the progress of the patient

Related Jobs

Communicate with the patient's family

Comfort the patient

Make sure the patient is not in pain

Maintain vitals on the patient

Emotional and Social Jobs

• Feel respected as a professional

• Feel like you're making a difference in the lives of your patients

• Be perceived as competent

Human and Environmental Constraints

• Patient and family are likely to be extremely stressed; could be angry and confused

• Patient and family may not remember instructions because of stress

• Must make sure that HIPPA regulations are being followed

that are *not* currently addressed at the retail clinic, however, deal with joint injuries and broken bones. This begs the question of whether a new platform could be established to treat orthopedic conditions on a low-cost, walk-in type of model. Based on this spectrum, this market would find such a solution to be very attractive.

The population that these data reflect would often be considered non-consumers for the health care system for this type of episodic care. When interviewing the workers for this study, I found that most of these conditions were those that they would not seek help for—they only accessed the health care system for extreme problems or for episodic problems for their children. For themselves, they would just let the illness run its course or do something only if it got extremely bad.

Overcoming the constraints is a vital part of any innovation initiative and often the most difficult to accomplish. In a recent study we conducted with physicians, some of the constraints included inconsistencies in reimbursement relative to the skill and experience required, the lack of support for wellness and preventative medicine, and the lack of patient responsibility for their own health and well-being.[3] These constraints are likely to be universal for a geographic area or a specific culture but must be localized as the innovation is diffused. Table 7-3 shows the ranking of constraints of physicians in an eastern U.S. state in 2011. There are interesting differences between PCPs and specialists; PCPs are more concerned with incentives being aligned, reimbursement for wellness, and new ways of treating patients (e.g., phone and e-mail), while the specialists are more concerned with the high costs of malpractice, regulatory compliance, and the amount of debt incurred to become a doctor.

Develop a Workable Solution

The most significant problem at the heart of the health care crisis is the system itself—a single business model and a single platform designed around a centuries-old method of practicing medicine—without a focus

**Table 7-3 Constraints Among PCPs and
Specialists in 2011 Study in the Eastern United States**

Constraints	PCP (N = 70) Rank	Specialty (N = 80) Rank	Constraint Type
The health care industry does not reward prevention and wellness	1	9	Human constraints: Physician
Patients do not take responsibility for their wellness; will not make lifestyle changes	2	2	Human constraints: Patients
Convenient methods of treating patients are not reimbursed, e.g., phone calls, e-mails, etc.	3	8	Human constraints: Physician
Patients often go to a facility that is more expensive than what is required to treat their condition, e.g., ER for a cold, etc.	4	6	Human constraints: Physician
The cost of training new providers is high, e.g., medical school, residency, etc.	5	7	Environmental constraints
Malpractice costs are high and cause physicians to practice "defensive medicine"	6	1	Environmental constraints
Medical practices operate on very small margins, e.g., no room for error, extra time with patients, extra help, etc.	7	4	Environmental constraints
Patients have an entitlement mentality	8	10	Human constraints: Patients
The medical community anticipates a shortage of key types of providers	9	11	Environmental constraints
Misaligned incentives cause inappropriate utilization	10	16	Environmental constraints
Patients want to sue/blame someone when something goes wrong	11	12	Human constraints: Patients
The amount of debt incurred by physicians for medical school is extremely high, resulting in high debt when starting their practice	12	3	Human constraints: Physician
The cost for medical practices for regulatory compliance is high	13	5	Environmental constraints

on the jobs to be done. In the health care system, the jobs to be done are highly diversified based on the extent of the medical condition, the difficulty with which the condition can be accurately diagnosed, and the skill required in treating the condition. Yet traditionally, and in many cases today, most of the myriad jobs within the health care system are addressed with the same platform and business model. The treatment of a sore throat, pink eye, or ear infection is conducted with the same infrastructure and business model as cancer, cardiovascular disease, and emergency medicine.

However, we are starting to see some positive changes and the evolution of new platforms and business models that are much more in tune to the job the patient is trying to get done. For episodic short-term medical problems, we are seeing the expansion of the retail medical clinic concept such as MinuteClinic as well as the beginning of telemedicine and virtual medicine. At the other end of the spectrum, highly specialized hospitals and solution shops are becoming more widely accepted, and new ones are being created within large general hospital systems. These new hospitals are extremely good at managing the entire condition the patient has taking care of all treatment for the condition within the same facility.

As shown in Table 7-4, matching the type of facility/platform to the type of medical problem is a significant step toward making the health care system more efficient. A single platform addressing such different jobs to be done provides some choice but to meet the highly complex issues and overserve the simple issues, or to focus on simple issues and fail to meet the needs of complex issues. Because the U.S. health care system is focused on the highest-quality medicine using state-of-the-art technology, the system by default will be highly overengineered for less complicated injuries and illnesses.

Successful innovation must be addressed job by job. By understanding the needs of the members at the point of service of these various jobs, solutions that make the most sense can be developed, using the most economical resources to achieve the optimal results.

Table 7-4 Effectiveness of Platform vs. Job Category

Highly specialized hospitals, Mayo Clinic, stroke centers, etc.	Overkill	Overkill	Optimal
General hospitals, multispecialty medical clinics	Overkill	Optimal	Insufficient
Retail clinics, full-service primary care clinics, telemedicine	Optimal	Insufficient	Insufficient
	Preventative medicine, easily diagnosed illness or injury, chronic condition management	Multifaceted illness or complex injury	Complex, diagnosed condition, long-term illness or extensive injury, terminal illnesses

Business Model

Vastly different jobs require vastly different business models.[4] This probably seems obvious, but it is not being done in the health care industry. Within the health care space, as the disease state becomes more clearly understood, precisely diagnosed, and treatment protocols well documented, the treatment can be conducted with much more efficient business models.

Clayton Christensen summarizes the core of the health care business model issue very well in *The Innovator's Prescription*: "The delivery of care has been frozen in two business models—the general hospital and the physician's practice—both of which were designed a century ago when almost all care was in the realm of intuitive medicine."[5] Intuitive medicine is defined as the highly specialized skills that physicians must use to diagnose a complex problem that does not have obvious evidence-based examples.

Chronic disease requires constant monitoring and working with the patients to ensure that they are following the treatment protocol and focusing on wellness—keeping the condition under control. Unfortunately, the current model of reimbursement for physicians who treat these conditions is exactly the opposite—the physician gets paid to see the patients when they are *not* well. Systems such as membership-based capitation, where physicians are paid a set fee for each "member" they are responsible for, ensures that the physicians are given an incentive to keep their patient base healthy, to be efficient in how they treat their patients and in how they educate the patients on maintaining their health and wellness. (See Table 7-5.)

Table 7-5 Evaluation of Different Forms of Health Care Delivery

	Basic Illness/Injury	Chronic Illness	Critical/Terminal Illness
Evolving platforms	Retail: Lower-level professionals, inexpensive facility, minimal equipment and diagnostics, rules-based treatment protocols	Disease management networks: Specialized groups based on disease state; use of specialists, and physician extenders (assistants, nurse practitioners, telemedicine, etc.)	Hospital and specialty care units; top-level professionals; state-of-the-art diagnostic and treatment equipment; facilities that allow the patient to obtain all necessary treatment
Revenue structure	Fee for service: Much lower rates since platform is less expensive	Fixed fee per person per year for management of the patient: Profitability increases with wellness	Major medical coverage through insurance using deductibles, coinsurance, and/or copays
Quality means	Get served quickly, low cost, available when needed; speed in recovery	Keep the condition under control; monitor in a way that is inexpensive and not a hassle	Best possible treatment to prevent death; maximum quality of life during the course of treatment
Key jobs of focus	Resolve the condition (injury or illness) Prevent the condition from getting worse	Ensure patients' adherence to treatment plan to ensure health and control of condition	Slow the progression of the condition Manage the symptoms of the condition

Different jobs measure quality differently based on the job the customer wants to get done and the method in which it is satisfied. Quality for the retail, therefore, is just as high as the advanced technology for critical illnesses since the definition of quality is to meet the needs of the customer. Because the value is measured differently, it is imperative that the business models reflect the definition of quality for each.

Thus retail medical clinics with their unique business model—inexpensive retail space, less trained providers of services, focusing on a limited set of health care issues—are a perfect match of the job to the business model.

Diffusion Considerations

Keep in mind that the jobs that are being uncovered are universal and stable over time. Regardless of which medical condition is being treated, the job to "obtain health care treatment" and its associated job steps will be the same whether the patient is seeking an appointment for an annual check-up or an appointment for open-heart surgery. The patient will still need to determine where to obtain care, determine who the best provider is for the condition, gather the necessary information to schedule the appointment, get transportation to the appointment, and so on. This is the beauty of using the jobs to be done with the detailed job map to explore the quality metrics on how well the job is being done. The ubiquitous nature of these jobs is what enables the diffusion of the innovation strategy and needs set around the world.

As shown in the stroke care example, any health care program will have a lot of data involved. The creation of universal job databases is designed to address this need and to share the job and need information across organizations and entities that are all trying to solve the same problem. As discussed earlier, the goal of social innovation is to make progress. Thus, if hundreds of organizations are exposed to the needs of people suffering strokes, then the likelihood of having improved stroke care across the country, and across the world, rises dramatically. When I taught at the Strategyn Institute, I would always ask my students, "If every company in your industry knew all of the unmet needs of your customers, who would win?" The answer? The company that is the most creative and can execute and get to market. As Seth Godin might say, "Whoever just does it and ships it!"[6]

Therefore, as discussed in Chapter 4, together with my clients and colleagues, we are working to create universal, global discovery needs archives (DNA) for the critical social innovation areas. Health care is most certainly one of these. Think about it. Strokes occur in every city in every country around the world. The treatment protocols are being solidified each year. The needs of the physicians, patients, families, nurses, all the ecosystem members are very much the same regardless of where the stroke takes place or where it is treated. The difference will be in the importance and satisfaction in the job steps executed when treating the stroke, the emotional jobs, and the constraints (especially the cultural, social, regulatory, technical, and political constraints) of how the stroke is treated in any given area.

Thus, once the job steps are obtained for ER physicians treating strokes, this set of needs can be used by hospital systems in the northeast and southwest United States, or any other country that treats strokes. If all hospitals that want to improve their ability to treat a stroke were to utilize the unmet need data and quantify them in their region to find out where they need to focus their idea generation efforts, the innovation in these spaces would evolve rapidly. Knowledge of the unmet needs and the constraints is the first step toward innovation. Working together, we can create these unmet needs datasets and catalyze innovation in the health care space. The health care DNA is the innovation equivalent of evidence-based medicine where documentation of what works organized by the condition (need) makes solving the problem much more effective.

In Summary

The health care system, while one of the country's most difficult wicked problems, offers great opportunity for new value creation. The key will be to focus on each of the job areas, overcoming the constraints to getting those jobs done, and creating platforms and business models optimized for the job. Using a structured approach makes this monstrous problem digestible, addressable, and ready for innovation. The use of a common framework creates a means for the sectors—nonprofit, corporations, and government agencies—to work together, thus accelerating the innovation process. The framework helps these diverse groups to focus on the problem from the same perspective and understanding, allowing each to contribute its areas of expertise to solving the problem. Such a collaborative approach can bring rapid development and deployment of a sustainable and affordable health care model.

Chapter | 8

RESOURCE CONSERVATION

The surge of population, the flattening of classes around the world, and the lifestyle that we have come to rely on requires a new look at our relationship with nature and the earth's resources. We can no longer afford to simply take from the earth without any thought as to how much is left. We cannot simply take from the earth and not give back. With such a mindset, the earth simply cannot sustain the projected 9 billion people that are expected to inhabit the earth by 2050.

Conservation traditionally has focused on preserving the environment and "protecting" it. The question is—from whom? Humans and nature must be able to coexist, each supporting the needs of the other. Peter Kareiva, chief scientist for The Nature Conservancy, has promoted this notion, even to the angst of traditional conservationists. He does not believe that conservation should be about just preserving the biodiversity of life, but instead, "The ultimate goal is better management of nature for *human benefit*" (emphasis added).[1] This "radical" approach suggests that successful conservation will take place when the needs of nature and the needs of humanity are *both* met. Instead of putting a protective barrier around nature, it brings people and nature together in a relationship in which the needs of both can be met. With the population of the world expected to grow by another 2 billion people, this is the right approach to sustain the levels of water, energy, and food that will be needed.

Water crises used to be limited to the developing world where the means and technology for storing large quantities of water did not exist. Today the developed world is experiencing severe draught and water shortages. Andrew Winston, noted author of *Green Recovery* and coauthor of *Green to Gold*, recently noted that water may become the next carbon and that many organizations are beginning to calculate their "water footprint."[2] Water, a valuable and ubiquitous resource, is often taken for granted—until it is no longer there. Before we get to that point, argues Winston, we must begin to manage it.

On the energy front, the race to produce clean energy is on and is currently monopolized by six countries that hold all the patents on today's clean energy technologies with Japan in the lead (via Hitachi Corporation).[3]

Then there's food. The price of food is skyrocketing because of low supplies compounded by crop damage resulting from flooding and drought, rise in population and prosperity, and the use of food products for biofuels.

"Going green is no longer a quaint 'nice to have' but has become a business mandate. Today's companies have little alternative but to embrace sustainability."[4] The key question is what is sustainability?

Define the Problem

In order to make progress on innovation for resource conservation, the first step is to get a solid problem definition including identification of the ecosystem members, the jobs they are trying to get done, and the constraints that are standing in the way of success. To simply say that companies must "go green" or become "sustainable" gives us little to go on. What is the actual job that is being executed?

"Green" and "sustainable" as terms are just too vague to work with and are solutions for jobs that companies, citizens, and governments want to get done. For example, some of the jobs around "green water usage" might include, "Manage the use of water resources in the home," and, "Reduce the amount of water used during commercial production." For energy, the jobs may include, "Manage the energy usage of the home/building," and, "Identify alternative energy sources." For food, the jobs may include, "Ensure adequate food supplies for current and future generations," and, "Distribute food to areas where food is scarce." The key is to zero in on the actual jobs that are to be done and then find a way to execute these sustainable jobs.

Consistent with this thinking is the article referenced earlier by the Nature Conservancy's Peter Kareiva: nature's needs must be considered alongside the needs of humanity. In our Social Innovation Blueprint, nature is simply another player in the scenario of, "Preserve natural resources"; it is a job executor in some contexts and a job beneficiary in others. The Social Innovation Blueprint in Figure 8-1 illustrates one program in which Nature is a job executor supporting the needs of people.

When we consider the needs of nature, we can approach them as we would for any other player in an ecosystem, with the exception that

Figure 8-1 Social Innovation Blueprint of Conserving Water

Social Innovation Blueprint	Challenge : _Conserving Water_	
Beneficiary Constraints		
Beneficiary **Animals and plants** Job : Obtain water for survival **People** Job : Obtain water for drinking and daily activities, e.g., bathing, washing clothes, etc.	**Animals and plants** Job : Obtain water for survival **People** Job : Determine alternative supplies of the resource Determine how to conserve what is left of the resource	**Animals and plants** Job : Obtain water for survival **People** Job : Adjust habits to conserve water Capture water for use
Status Quo	Shortage of Resource	Postshortage
Executor (s) **Nature's water systems** Job : Avoid depletion of water through changes made to the ecosystem, e.g., deforestation, diverting rivers, etc. Maintain water tables	**Nature's water systems** Job : Preserve water during shortage Minimize damage to wildlife/plants **Supporting city/state water management groups** Job : Develop plans to replenish water Determine how many people can be served by existing water resources	**Nature's water systems** Job : Restore/adjust to the new ecosystem Replenish water storage **Supporting conservation groups** Job : Educate people on water conservation Product land from over-development
Environmental Constraints		
Bold indicates primary beneficiary and primary executor	_Job is in italics_	

we can't directly ask nature about its needs. However, science over the last few centuries has provided significant insight into the needs of nature. We know that soil must have time to recover after it produces crops, that aquifers must maintain certain levels to be efficient, that the ecosystems and habitats must be maintained, and so on. The knowledge is there. We must now acknowledge that these are real needs of nature.

Identify Unmet Needs

In 2008, my colleagues and I conducted a study for the Continental Auto-mated Building Association on what homeowners believed "going green" meant. We found that this simple phrase included nearly 100 different jobs, with energy conservation accounting for only a handful of jobs which were ranked at the bottom of the first quartile.[5] Jobs that ranked higher than energy conservation included protection from toxins in the air, water, and food, as well as reduced waste in landfills. So while green is often referred to with respect to energy conservation, for consumers this is just one small piece. As shown in the opportunity spectrum in Figure 8-2, items pertaining to food and water safety rated higher than energy con-servation items. Also some of the highest opportunities within the energy group were focused on measuring energy usage.

Figure 8-2 Opportunity Spectrum for "Going Green" Jobs in 2008

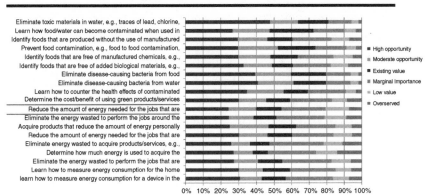

Interestingly, while the health care data yielded several significant opportunities with 30–35 percent of the market in the high opportunity range, the conservation data are even more extreme with nearly a dozen jobs at the 40–48 percent high opportunity mark, indicating that this issue has even more urgency for innovation. The difference between the two is that over the past couple of decades, the health care system has improved to the point where it has a large population that is "satisfied." However, in the realm of conservation and sustainability, there is only a small percentage who are satisfied—most are very or extremely dissatisfied.

Why would *energy* conservation rate so low in the green list of jobs? As pointed out by Clayton Christensen and his colleagues, "The developed world's existing energy infrastructure is so cheap and convenient that it creates large barriers to adopting new energy technologies."[6] On the other hand, there are scares over the contamination of food and water regularly in the news and very few products on the market that actually help to address these jobs. Therefore, given the breadth of jobs in the conservation space, it is imperative that the importance and satisfaction of all of the jobs be understood so that the focus of innovation efforts is on target.

This notion becomes apparent when we consider the slow adoption of solar energy and wind power. In the developed world, the ubiquitous availability of energy and its affordability make it extremely difficult for alternative energies to compete. Solar and wind energy are competing directly with traditional energy programs, and companies cannot make the new technology cost-efficient enough for customers to switch from their existing energy source. Only those customers who place high importance on jobs related to protecting the earth's natural resources, avoiding damage to the world's climate, and other jobs that are focused on maintaining a healthy planet will pay more to purchase more expensive renewable energy.[7]

The highest opportunities within the energy conservation (and water) relate to being able to measure the amount of the resource used. There is significant frustration with the lack of actionable information about usage, making it very difficult to understand consumption rates and the effect people's actions have on their consumption. For instance, if people leave the water running while they're brushing their teeth for the standard two minutes, it is estimated that 12 gallons of water are used.[8] Does the average consumer know this? How much additional energy is used if the air conditioner is turned down by two degrees in the summertime? How much water is used for every extra minute spent in the shower? Without visual feedback, people feel helpless to control their resource usage. According to the study, they want the feedback as they are going along so they can make appropriate decisions.

Develop a Workable Solution

In order to create a workable solution, we must find the synergistic needs of the primary members of the ecosystem. In the conservation arena, there are four key groups (although there are over a dozen members in the complete ecosystem). These four entities must coexist in a way that allows each to achieve its primary jobs with minimal negative impact on the jobs of the other members. Also, while this may seem obvious, the creators of new technology must be sure they are creating solutions for jobs that the customers are *already* trying to do. There are many technological solutions for homes and commercial buildings that are available today; however, for a large portion of the population, these do not address a job that is important. Until the *importance* of the jobs around energy conservation and renewable resources increases, adoption will continue to be slow. It is imperative that we work to move the importance needle.

These renewable energy sources, while not yet taking hold in developing countries, should be targeted at nonconsumers, primarily in the rural areas of *developing* countries. In these areas, the new energy sources would be very welcome for two key reasons: (1) they are better than the current alternative—no energy at all, and (2) there is no existing infrastructure that would need to be overhauled.

If you recall from earlier chapters, when one member of the ecosystem attempts to satisfy a need at the expense of another, there is likely to be failure. The state of California attempted to create quotas for electric vehicles in the early 1990s which would meet a need of the government (reduce gasoline consumption, reduce pollution, etc.) and the people (reduce the cost of transportation, reduce pollution, etc.). However, at the time the infrastructure, materials, and technology were so expensive that the automakers would not be able to meet one of their key needs (achieve a desired profit on each vehicle sold). As expected this proposal failed. Innovations in policy or programs that create a substantial burden on any of the other members of the ecosystem are likely to be met with strong resistance.[9] If the government begins to understand the needs of the other members of the ecosystem first, before attempting to impose regulations and legislation, there is a much higher chance of success and an opportunity to avoid so much wasted time and effort.

Contrast this situation with the regulation to replace lightbulbs with compact fluorescent lightbulbs (CFLs). The mandate was a win-win in that the industry found the margins to be much greater with the CFLs and the homeowner would see significant energy cost savings. By addressing key needs of both constituents, the regulation was welcomed instead of rejected. It is vital to remember that where industry loses, backlash is very likely to derail progress.

The human and environmental constraints must also be addressed in order to ensure that any solutions created become viable. For instance, adoption of new energy sources met with a significant environmental constraint—a lack of transmission infrastructure to distribute the energy. "Oftentimes, large solar plants or wind farms are built in remote low-population areas. Massive investment is then needed to construct the transmission lines necessary to carry the renewable energy to high-population areas."[10] To address this problem, a synergistic solution was launched in July 2011. The U.S. Federal Energy Regulatory Commission passed a new planning law which requires that a plan be developed by many regional players, allowing the high cost of the infrastructure to be distributed, instead of one player having to take on the burden of providing a single transmission line. A quote from the American Wind Energy Association sees this as a major game changer that can massively accelerate the growth of the infrastructure. Thus the benefits are clear to the energy producers. As for the consumers (both residential and commercial), they benefit "by opening the grid and allowing regions to access lower-cost energy from further away . . . [which] will reduce the power of outages and rolling blackouts."[11]

The synergistic needs of the ecosystem members must be met in order to drive successful innovations. With the Nature Conservancy's focus on meeting the needs of both humans and nature, it has seen much success; two of the cases are summarized below.[12]

Human need: Food

Nature's need: Avoid overfishing

In Micronesia, TNC has created a network of local fishermen who maintain networks of marine areas that are protected. These fishermen fish to feed their families but are also responsible for protecting these areas from poachers, thus preserving the sustainability of fishing in the area.

Human need: Water

Nature's need: Avoid depletion of water through deforestation

TNC created local programs to collect fees for water usage from the locals and uses the funds to restore the watershed ensuring that there will be water available for the people in the future.[13]

The key to the success of this program is a focus on the needs of the local people, the people whose land is in danger. Once those needs are identified, teams can then work together to find ways that the needs of humankind and the needs of nature can both be met in a symbiotic relationship that will ensure a sustainable environment for generations to come. Trade-offs must be evaluated and compromises considered. For example, "The goal of nature-friendly commercial agriculture could be reached through a combination of modifying farming practices, creating wildlife corridors through agricultural lands, breeding or engineering plants that require less water, are more nutritious and produce their own pesticides."[14]

In addressing the need of reducing the waste that goes into the landfill, there are also some interesting innovations being undertaken. Austin, Texas, has created a major campaign called Zero Waste with a goal of *reducing landfill waste by 90 percent by 2040*, including the production of a reality show where four families have five weeks to reduce their trash to zero. This campaign is designed to give residents ideas on how to reduce their waste and educate them on the need for zero waste in an entertaining platform that the entire family can enjoy. This is an example of an innovative way of moving the importance needle and getting people the information they need in order for them to take action.

Business Model Considerations

As in the other areas of social innovation, business model innovation is necessary in order for green technologies, services, or programs to succeed. Because of the intricacies of the ecosystem involved in resource conservation, it is imperative to involve all sectors—government, nonprofits, and corporations. Without the input and efforts of all three entities, innovations in sustainability have little chance of success.

Within the business model canvas, resource conservation has a quite literal element of the environment and society.

The Nature Conservancy recently created a financial investment tool called the Conservation Water Trust Fund and launched it in Colombia. While this is technically an investment tool, it has an immediate impact on the environment and in changing behaviors at the local level. The fund "receives contributions from Bogotá's water treatment facilities to subsidize conservation projects—from strengthening protected areas to creating incentives for ecologically sustainable cattle ranching—that will keep sedimentation and runoff out of the region's rivers."[15] Why would they participate? Instead of spending millions to remove pollutants, they are investing in conservation up front.

The Trust Fund then distributes money and resources to help locals learn about forest conservation, financing changes to ranching practices, and helping farmers update their equipment and sometimes even modify their crops to drastically lower water usage. This first-of-its-kind fund is working actively with local farmers, businesses, scientists, conservationists, community members, and many others to produce a sustainable environment thereby preventing disastrous water depletion.

The extent to which the conservancy has involved all the major ecosystem members is key to its success. It has involved the local people and the business owners of large and local businesses; it has directors on the ground to coordinate the program. "The Conservancy brought together a broad range of public and private stakeholders—many of which had never collaborated before—to . . . broker the landmark agreement."[16] It is projected that the *savings to the water treatment* facilities will exceed $4 million per year. The local farmers and ranchers benefit by receiving grants that help them switch to sustainable operations in exchange for commitment to long-term conservation agreements. The funds can help ranchers purchase higher-quality cattle, resulting in more milk from fewer cows, which results in less grazing fields being created from deforested land.

According to Aurelio Ramos, director of the Conservancy's Northern Tropical Andes Program, "In financial terms and in actual conservation results, these funds yield a tremendous return on investment."[17]

Figure 8-3 shows how this trust fund plays out in the Shared Value Framework. It is clear to see the significance of the value proposition given the attempt to focus on meeting needs of all players involved.

Figure 8-3 Social Business Model Canvas for Water Trust Fund

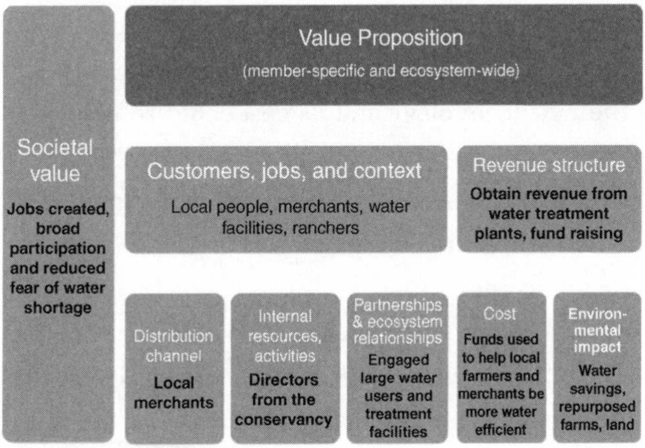

Diffusion Considerations

As stated earlier, the adoption of renewable energy sources has been slow and discouraging, largely because of the inefficiency of our existing infrastructure. The developing world is likely to adopt these new technologies much more readily because it has large populations of nonconsumers.

In the developed world, green technologies will need to be enhanced to meet the jobs to be done for the customers in these areas. For this to happen, one of two things must take place. First, the *importance* of the jobs around maintaining the health of the planet could increase dramatically through continued drastic global climate change, a significant depletion of natural resources (such as being cut off from foreign oil), or other natural or human-made events that would cause people to care more deeply about the natural resources we have and their use.

We've seen this in areas of severe drought where the importance of preserving water increases dramatically as lakes are drained, water tables depleted, and uncertainty about the future takes hold. We saw this in the study on energy conservation mentioned above, where those customers who had higher monthly energy bills rated nearly all the needs as much more important and were far less satisfied than those with smaller monthly bills.[18] Some of the key items that became increasingly important were being able to set operating parameters of energy-efficient products as well as how to determine if they are operating at their optimal levels, learning how much energy is actually needed to control the climate within the home, and identifying products that could reduce the overall consumption of energy.

Satisfaction with the new technology and its cost could substantially increase. As manufacturers become more efficient at producing these new technologies and the prices become more in line with what traditional electricity costs, customers will find that the new solutions can better satisfy some of their jobs or outcomes.

For instance, if traditional energy sources began to skyrocket in price, the overall cost of these new solutions would eventually become equivalent to that of traditional energy, thus increasing the adoption of the new technologies.

Additionally, if the new technologies include devices that were to allow the customer to get other jobs done as well as providing energy for the home, such as storing excess energy to give back to the grid, providing a means to cycle energy during peak times and generating revenue for energy given back, then the relative attractiveness of the new technologies might increase.

Another area that requires significant attention on diffusion is that of waste management. Many large companies have moved to a zero waste policy, and this is fortunately catching on. However, to make significant improvements will require innovation in packaging, recycling, and distribution. A quick Internet search will reveal how many new products exist that are completely biodegradable and made from renewable resources— right down to the bags, to-go boxes, and cups used at the millions of fast food restaurants. The technology is there; now it is a matter of moving the importance needle so that companies and consumers will adopt and even demand these low-waste solutions.

In Summary

The issues we face as residents of a planet are significant and widespread. For too long we have not viewed nature as a necessary part of the commercial ecosystem whose needs must be taken into consideration. Fortunately, many organizations have been pushing for a long time, and this topic is now mainstream and no longer reserved for the fringes and the "fanatics." It is a human problem and must be addressed. I firmly believe that the talent of innovators around the world can solve this problem; they can create new energy sources, develop zero waste plants, develop new solutions to help consumers manage their resource usage, and much more. The key is getting the incentives in place to do so and for companies to come to the party and bring their talent, patents, and technology with them. We need government to step up and create the policies that are necessary to overcome the resistance and complacency we have resulting from the efficient system of energy and water distribution we have today. It will take not only a village to solve this problem, but a world.

Chapter | 9

WHAT CITIZENS WANT

The Social Impact Framework illustrated throughout this book has no greater purpose than that of bringing innovation techniques and the philosophy of collective impact to every part of our government. President Obama in 2011 identified innovation as a vital and core value of this country; however, the goal of innovation will not be fulfilled without a structured process and the cooperation and collaboration of the public, private, and nonprofit sectors. It is the government's role to solve some of the world's most wicked issues: climate change, poverty, a broken education system, nearly bankrupt social services, substantial unemployment, and the continued escalation of our national debt. However, it is clear that the government alone cannot solve these problems. It will take the collective impact of all three sectors—government, nonprofit, and corporations—to make inroads on any of these problems. Satish Nambisan, associate professor at Rensselaer Lally School of Management and Technology, sums it up well: "The performance of American government in the 21st century will be shaped by how well it adopts collaborative innovation to harness external resources and creativity in addressing the nation's most challenging issues"[1] it is imperative that the three sectors share a common language and a common framework for developing innovative solutions.

The actual job of the politician is to address the needs of the citizens within the boundaries of their district or their sphere of influence. Therefore, it is imperative that politicians *know* the needs of their constituents and the prioritization of those needs. However, instead of asking the citizens what their needs are, the politicians typically put forth their plans and ideas and conduct polls to judge the value of these ideas. The problem is that the polls are asking the wrong questions. The primary question should be, "What do citizens want?" If the needs of the citizen base are known, then politicians can develop valuable ideas to begin with, eliminating the trial and error process that exists today.

Just as successful corporations have done, governments must learn *how to listen* to the customer (the citizen) and put what the citizens want in a common language (jobs to be done)—a language that everyone (the citizen, the rocket scientist at NASA, the hot dog vendor on the street, and people on Capitol Hill) will understand. It is at this point that real conversations for change can happen and the government can begin to develop the solutions that will matter to the people.

While the sheer number of jobs involved in the government sector, from the city, state, and federal perspective, may seem overwhelming, the good news is that each job has to be studied only once. The government sector can make the best use of the discovery needs archive (DNA) since cities and states are not competitors and can easily share information. At the end of the chapter, we discuss the use of this DNA as a means of rapid deployment of citizen inputs and nationwide sharing of innovative outputs. This DNA system can play a vital role in creating President Obama's vision to, "Out-innovate out-educate, and out-build the rest of the world. . . . and make America the best place on earth to do business."

Because of the expansive number of government scenarios, instead of being able to go in depth on a single scenario, this chapter discusses several scenarios at a higher level to provide a sense of how to apply the Social Impact Framework to these government sectors. Further whitepapers will be developed on these important programs and posted on our website www.TheSocialInnovationImperative.com.

Education

The education system in the United States has evaded large-scale reform despite the efforts of nonprofits, government mandates, and dedicated teachers and education professionals. Much of the failure has been attributed to insufficient budgets, a monopoly style incumbent, and goals that have become a moving target. It is precisely because of these circumstances that a clear understanding and prioritization of the needs of all parties in the ecosystem is vital for success.

Within the education scenario, there are several members of the ecosystem (teachers, students, school administrators, parents, librarians, etc.) who are all working toward the goal of producing capable, knowledgeable citizens. Because of the rapid pace at which technology is advancing and our world is changing, the education system must continually adapt to the changing needs of the country, producing students that are ready and able to do the jobs of tomorrow, not the jobs of yesterday. If we are not producing students who have the skills needed to be effective for the jobs that are in demand when they graduate, then the education system has failed. It is therefore imperative that we understand not only the needs of the students and parents, but also the needs of their future bosses—employers, higher education institutions, government agencies, and the like.

There are four elements that are critical for success moving forward: cross-sector collaboration, common language and process, disruptive new platforms, and new job executors. We see all of these factors in play within programs that are starting to see real progress.

In Cincinnati, a group called Strive achieved a vast improvement in local education and attributes its success to the "commitment of a group of important actors from different sectors to a common agenda for solving a specific social problem."[2] The researchers note that innovation in

Figure 9-1 Social Innovation Blueprint for Education

Social Innovation Blueprint		*Challenge: Education in low income areas*	
Beneficiary Constraints			
	Multi-tasking,multi-media raised children are easily bored	Children do not see the connection of these tests to their life	
Beneficiary	**Student** *Job: Learn content*	**Student** *Job: Complete the standardized test* State Education Departments *Assess how well students are learning material; compare schools' competence*	**Student** *Job: Use scores to determine what areas need further study*
	Status Quo	**Crisis period** Standardized Tests	**Post-Crisis**
Executor(s)	**Teachers** *Job: Communicate content to student; Ensure students learn material* Parents: *Provide encouragement to the student; provide help/support to the student*	**Teachers** *Job: Conduct the standardized test* Parents: *Ensure students are prepared for the test*	**Teachers** *Job: Justify/support scores, adjust lesson plans to improve*
		Often either both parents working or a single parent household	
	Environmental Constraints		Teachers are often over-loaded–too many students per teacher

Bold indicates primary beneficiary and primary executor   *Job is listed for all parties in italics*

education cannot be mandated by the government or driven by a single, even very large and powerful nonprofit. It requires the collective work of an entire community.

Jeff Edmonson, the program's director, attributes the success of this program to its having a *common language and using data to drive decisions.* Essentially, they have implemented Six Sigma approaches to continuous innovation.[3] Applying methods such as jobs-to-be-done and the Social Impact Framework can provide further discipline and common language around the needs of the members.

Given the problems with the existing platform of the education system today, we must find new and disruptive means of educating students. The challenge is that the education system is highly monopolistic; it is a singular vast system. It is highly unlikely, if not impossible, that disruptive innovation will come from the incumbent players within the monopoly. The movement for disruption is most likely going to have to come through entrepreneurs targeting home-schoolers, parents, and students themselves who want to supplement the learning of the education system. Another avenue includes alternative type schools, such as charter schools, that have more flexibility and freedom within their curriculum development.

The job executor is also likely to shift from the teacher to the student and the parents. Much of the content that children can access today far exceeds what is taught in the classroom and is often presented in a style that is more conducive to learning. For example, sites such as FunBrain.com, Learn4Good.com, Jumpstart.com, and literally hundreds of others, help kids learn everything from math and reading to complex physics notions; all of these sites use interactive games, fun graphics, and entertainment as the means by which to communicate with the kids, engage them, and help them learn the content. The sheer number of educational options on the Web is astounding and illustrates the success of using game-style content and platforms to help kids learn. With such vast and engaging content available in a format that kids enjoy, the future classroom, textbook, and teachers should all be reconsidered to create an environment that allows children to learn in the most optimal way. In the schools of the future, it is likely that teachers will play a different role, more of a coach and navigator of content, a facilitator of dialogue, and resource for explanation.

An example of the disconnect between the focus of school curricula and what the country needs is evident when we look at the topic of innovation. President Obama, in 2011, issued a quest and vision for the United States to regain its position as the leading country in innovation. However, very little time, if any, is spent teaching children what innovation is, how to look at the world in different ways (such as Edward DeBono's Six Thinking Hats), how to combine things in unique ways, and simple basic concepts of creativity. By second or third grade when the crayons are put away, so is the creative thinking; it is replaced with a focus on memorization, rote learning, and following rules. While these techniques may benefit the needs of the *teachers* to maintain a disciplined class, or for school districts to be able to compare performance of one school against another, these techniques do not meet the current needs of the targeted customer—the students and their future employers. To improve our education system, these students, parents, and employers that will hire the students must be the focus of everything the school system does. Instead, school systems today are often "top down" with the curriculum coming from state or federal guidelines geared more at protecting the status quo and passing standardized tests than creating the next generation of entrepreneurs, innovators, and leaders.

Military Applications

There are literally dozens of applications of the Social Impact Framework in the military, including creation of new technologies to support the troops in the field and creation of new programs and services to assist soldiers in transition to civilian life. There are even applications to help military personnel understand the needs of the people whose country they are fighting in so as to gain their trust and support. One of my favorite examples of the latter is the CIA's distribution of Viagra to chieftains whose help they needed in fighting the Taliban in Afghanistan. "Whatever it takes to make friends and influence people—whether it's building a school or handing out Viagra," said one longtime agency operative and veteran of several Afghanistan tours.[4] Even in conflicts that involve life and death situations, solutions can be found by paying attention to and understanding the needs of the people involved in the ecosystem.

I've seen this first-hand as both the proud mother of a U.S. Marine and as the daughter of a Colonel in the Air Force. Thirty years ago the family of the soldier was not given much consideration; they were expected to fall in line, pick up and move without a second thought, and do everything they could to support the service member. In today's military, more attention is paid to the needs of the soldier's family. The Marine Corps has done an amazing job communicating with the family throughout my son's three years of service. From the parents day at the graduation ceremonies to the "Jane Wayne" days where the military wives and families learn about the equipment, shoot an M-16, and run the obstacle course, to the dedicated communication officer assigned to every deployment—the military ensures that the family is kept informed at every stage. They have taken the

Figure 9-2 Social Innovation Blueprint—Military Example

Beneficiary constraints		Fear and uncertainty for both soldier and family	Stress of reintegration in noncombat zone for both soldier and spouse
Beneficiary	**Soldier** *Conduct operations on base* **Soldier's spouse/children** *Job: Support the soldier*	**Soldier** *Conduct missions at the deployed location* **Soldier's spouse/children** *Job: Support the soldier while on deployment*	**Soldier** *Reintegrate into noncombat life* **Soldier's spouse/children** *Job: Help the soldier reintegrate back home*
	Status quo—on base	Deployment	Post-deployment
Executor(s)	**Marine unit** *Job: Outline operations to be conducted on base*	**Marine unit** *Job: Identify and distribute missions while in the field; protect marines from harm* Supporting executors: Translators *Job: Assist the soldiers in executing the mission*	**Marine unit** *Job: Debrief marines and prepare them for return from combat* Supporting executors: Health and wellness services *Job: Ensure the soldier's health and well-being*
		Inadequate communication with the soldier during deployment	Not enough mental health professionals
Environmental constraints			

mystery of what the life of the service person is like and made it transparent to those whose lives are affected by the service person's work. This is an excellent strategy for the military to maintain the size of their all-volunteer force mostly by keeping attrition low. Once they've spent the money training the soldier, they want to keep the soldier. They have discovered that if they meet the *needs* of not only the soldier, but his/her family, then the soldier is happier, more stable, and more excited about the job.

To illustrate the benefits of the Social Impact Framework, Figure 9-2 is an example program for soldiers and their families. By understanding their needs during periods when they are stateside, deployed, and postdeployment, the military will have more concrete information on what kind of support these soldiers and families need. This vital understanding can be valuable in avoiding significant problems faced by returning soldiers such as suicide, alcoholism, and the like.

The blueprint shows a preliminary framing of such a program. Breaking it down further, an example of a "needs table" is shown in Table 9-1 for the

**Table 9-1   Needs Table for Preparation
for Deployment of Soldier by Spouse**

Job Map Step Name	Job Step
Focal Job: Prepare for Deployment	
Member of Ecosystem: Military Spouse	
Define	• Determine the scope of the deployment, e.g., length of time, location, etc.
	• Determine when the deployment will begin
Locate/prepare	• Help the soldier obtain supplies needed for the deployment
	• Take care of paperwork for the deployment
	• Ensure understanding of what activities/tasks the spouse must take on during the deployment that the soldier used to do, e.g., manage finances, maintain the yard, etc.
Confirm	• Ensure that the soldiers have everything they need to deploy
	• Ensure that all paperwork is submitted appropriately
Execute	• See the soldier off
	• Obtain notification that the soldier has arrived in the deployment area
	• Notify friends and family of the soldier's status
Monitor	• Determine how the soldier is doing
	• Boost the morale of the soldier, e.g., letters, care packages, etc.
	• Notify the family of the soldier's status
Modify/conclude	• Prepare for the return from deployment
	• Welcome the soldier home
	• Incorporate the soldier back into family life
	• Ensure that the soldier is not suffering from any problems resulting from the deployment, e.g., depression, anxiety, physical ailments, etc.

Related Jobs

• Connect with other soldiers' wives
• Take care of one's health and well-being while the soldier is gone
• Get help to care for children while soldier is gone.

Emotional and Social Jobs

• Feel confident in being able to manage the household during the deployment
• Feel comfortable about being able to get information as needed
• Feel secure that the spouse is well trained for the mission
• Be perceived as a good soldier's spouse

(continued)

Table 9-1 (*continued*)

• Be perceived as competent
• Be perceived as handling the situation well

Human and Environmental Constraints

• Spouse and family are likely to be extremely stressed, could be angry, sad, etc.
• Spouse must keep the spirits of the soldier up while dealing with his or her own sadness and stress
• Big events can often happen during deployment such as the death of a loved one, birth of a child, etc.

job of "prepare for deployment" for the military spouse. While this study has not yet been completed, it is one that would yield valuable information for our troops as the two conflicts in the Middle East wind down.

By working proactively to identify potential issues for the soliders and their families, the military will have better insight into the types of solutions that are needed to make the deployment process more manageable. Once the qualitative work is done, the data can be quantified by branch to identify differences among the various branches of the military, by bases or posts to get further detail on a specific sector, and even within specific battalions. This type of work is not a "nice to have"—it is a "must have" in order to support our troops in today's wars and the wars to come.

Disaster Management

Another example of social innovation that government is usually and ultimately responsible for is that of disaster management and disaster mitigation. The response to disasters has long lacked significant innovation and progress. According to a recent article in *Stanford Social Innovation Review*, the government grossly underspends in prevention of disaster even though, "For every dollar the U.S. federal government spends on disaster mitigation, it saves anywhere from $8 to $15 in spending on future relief" of those disasters.[5]

Whether disaster relief and/or disaster prevention, both require a significant level of collaboration among a very large ecosystem—the federal

Figure 9-3 Social Innovation Blueprint for Disaster Response

	Status quo–disaster prevention	Disaster period	Post-disaster
Human constraints	Sense of invincibility—"won't happen here" mentality—therefore, citizen preparations are weak at best	Panic, chaos, confusion	Frustration, disbelief, denial, depression
Beneficiary	**Citizens** — Job(s): *Prepare for the possibility of a natural disaster* — *Maintain awareness of pending disasters*	**Victim's family** — Job: *Determine the current health status of the victim and or whereabouts* — **Victim** — Job: *Obtain assistance for oneself* — *Help others in the affected area*	**Victim's family** — Job: *Assist the victim in finding medical care transportation, a place to stay* — **Victim** — Job: *Recover from injuries, recover/restore physical property, find intermediate shelter if needed*
Executor(s)	**First responders** — Job: *Train and prepare for response to natural disaster; create response plans* — Supporting executor: Red Cross — Job: *Provide training and preparedness advice to citizens on what to do in a disaster*	**First responders** — Job: *Provide emergency aid to victims at the disaster site; coordinate activities among teams* — Supporting executor: Aid Agencies — Job: *Provide assistannce to the first responders; coordinate information with victims' families* — Supporting executor: Hospital ERs — Job: *Provide emergency medical care*	**Red Cross (aid agencies)** — Job: *Help victims apply for needed aid* — **Hospitals** — Job: *Continue providing medical care* — **Mental health workers** — Job: *Provide mental health service to victims and families* — **FEMA & private insurance cos** — Job: *Assess extent of damage process claims for victims*
Environmental Constraints	Communication to rural areas may be impaired	Damage likely to infrastructure, roads, bridges, etc. making it difficult to get to victims — Very likely that there will be more injured than can be handled by the hospital system	Increase in construction work; may exceed workforce in the area — Need for temporary housing for large numbers of people

government; state, county, and city governments; nonprofit aid agencies; local law enforcement; first responders; local hospital systems; and so on. Especially important in the disaster relief issues are the constraints—both human and physical—which are substantial in this scenario.

The initial blueprint for this scenario has been mapped and is shown in Figure 9-3. The jobs and constraints for each of the areas need to be captured and then prioritized by the various audiences, after which we will be able to identify synergistic and conflicting needs. Once the needs of the ecosystem members are obtained, then we will have a model for continuous innovation for new programs, products, and services that will ensure effective and efficient disaster response.

The value of such a program is that the results of the program can be shared with all city, county, and state emergency management groups; the data can be localized for each area, and solutions can be shared via the DNA collaboration database. With the increases in natural disasters we've seen over the past couple of years—record flooding, killer tornadoes, wildfires engulfing several states—we cannot afford to wait to develop solutions that will make responding to these disasters more successful. We must start now.

Messaging and Budget Prioritization

Getting the message right is especially important in the government sector. Whether it is a mayor speaking to city council members or a presidential candidate talking to potential voters, the time spent on crafting the message should be based on the needs of the constituents the politician is speaking to. By focusing on the jobs that the citizens want to get done, the politician can move from large vague terms such as "economic growth" to job-based terms such as "develop a plan for new job creation," "identify new areas of innovation that the state can participate in," "identify areas in which the United States is losing business to other economies," and so on.

Let's look at an example of a candidate's platform on education as described in the *Boston Herald*:

> *Education:* Supports a new school construction program to improve crumbling schools. Supports recruitment of a "new generation" of teachers, improving teacher pay, and improvement in early childhood education. Opposes school vouchers. Has called for a "STEP UP" summer learning program for disadvantaged children through partnerships between community groups and schools. . . . Supports merit pay for "master teachers" but opposes merit pay for teachers based on test outcomes.[6]

While this platform outlines a lot of ideas and solutions, for the average American who has not studied all these programs, it does little to explain how exactly this platform will help, what problems it will solve that are valuable to the parents, students, and teachers. Table 9-2 takes these platform elements and illustrates how this messaging could be vastly improved by identifying the needs that the politician is seeking to satisfy for the target audience.

Table 9-2 Turning Messaging Elements into Job Statements

Messaging Element	Parents—Job Statements	Teachers—Job Statements	Students—Job Statements
New generation of teachers *(vague statement)*	Not clear. What does this mean to the parent? Why is this valuable? What will be improved?	Not clear. Change the role of teachers? Change the type of teachers hired?	Not clear. Will this change how students learn? What is the impact on students?
Improve teacher pay *(good job statement but needs to be broken down further—see examples to the right)*	Prevent teacher turnover resulting from low wages Attract good teachers to the profession	Obtain fair pay Obtain pay that is consistent with the schooling required Obtain pay that is commensurate with experience, etc.	Keep good teachers Have teachers that seem to like their jobs
Improvement in early childhood education *(OK—but needs to be put into job statements as shown to the right)*	Ensure that children are ready for school upon entering the public system	Ensure that children have the necessary skills before entering the public school system	Avoid feeling overwhelmed upon entering school; feel confident that I am ready for the classroom

The key challenge within the government sector is deselecting. Of course all these things are good to have, but we cannot afford all of them. This is where a solid understanding of the *prioritization* of the jobs is valuable. If a governor, for instance, wants to create an education program that will fit within the budget, the first piece of information that should be obtained is the prioritized list of needs of the students—the primary beneficiary of the education system. Second, the governor needs to understand what needs are currently unmet that are synergistic among parents, teachers, and students. By focusing in these areas, the education program will be certain to hit the critical components of value for the ecosystem members.

The hidden benefit here is key—the data will also show what needs are *already met well enough and need no further help* (existing value needs, and those that are already overserved or have low value). Nine times out of ten when working with clients, we find at least one or two projects or initiatives that are actively being worked on that fall in these categories when the needs are quantified. Companies are spending millions if not billions of dollars on programs that have no real value to the target audience. So, when trying to create an education program, the politicians can be armed with the most important needs to create new solutions, while at the same time weeding out programs that are targeting needs that are not valued or are already well met. Without having a prioritized list of jobs for all members involved, politicians will continue to be guessing at what should be cut and what should be given additional funding. By having solid data to support these decisions, the process becomes transparent to teachers and parents and is an objective decision.

Discovery Needs Archive

While it may not be common practice today, it makes sense that all the departments and agencies of the government have a vested interest in *sharing ideas*. Think about it. All of the cities across the country have the same set of needs that their constituents are trying to get accomplished; what is different from one city to the next is which needs are important and which are well satisfied. But the core set of need statements is the same.

So, once a city identifies which needs are unmet among their population, they begin working on solving the problem. The first thing that should be done, however, is to ascertain whether another city has already encountered this need and has already created a solution for it. A database, such as the discovery needs archive (DNA) that we've referred to throughout the book, is a vital tool for sharing innovations across the country, matching them to unmet needs, and knowing how effective they are.

The example in Figure 9-4 shows a new city that has just completed its quantitative sample of city job needs and has found that in the rural areas the residents are having a hard time getting alerts or notifications of pending dangerous weather. After checking the DNA, it can be seen that

Figure 9-4 Discovery Needs Archive

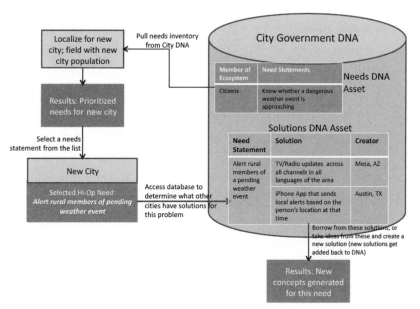

of the City of Austin (Texas) created an iPhone app that provides these notifications. Jacksonville's team can now contact Austin's disaster management department to get more information on the idea. Ideally this type of system could also include a star rating system and an opportunity for other cities to provide reviews and feedback on the ideas they have tried. This is the kind of sharing that will make our country an innovation powerhouse. Instead of re-creating the wheel time and again, often uncertain whether the wheel is even needed, we can finally approach innovation armed with statistically valid data, and work in cooperation with other similar organizations around the country, and even involve the citizens to help develop on-target solutions.

When a database with such a wealth of information is coupled with citizen-fed ideas, social innovation can take place at a remarkable pace. This new wave of citizen-engaged innovation has begun to yield excellent results. A group of researchers at the University of Washington were at a complete loss in their effort to determine the structure of a protein that causes AIDS. The problem was very complex as the elements they were investigating were smaller than that which could be detected visually, even with a microscope, and could be put together into so many permutations that even advanced computers could not get through them all. After spending 14 years using advanced imaging techniques and powerful computers they determined to engage a larger, unique group of people to solve the problem—gamers. They created a video game that would allow players to attempt to design the most accurately folded proteins—a lot like a 3-D jigsaw puzzle.* The game was set up to be engaging and competitive, and the players knew they were working on something important. With this combination of platform and a large group of determined people, the problem was solved in *ten days!* This is a fantastic example of the resource principles we discussed earlier—have someone else do the work for you. These so-called *open innovation* networks are becoming increasingly popular in the corporate world as well as with city, state, and even federal agencies that have begun to work *with* their citizens to solve problems. However, when asking citizens for solutions, you want to be very sure you give a specific, detailed description of the problem you are asking them to solve.

The Social Impact Framework makes Open Innovation networks even more powerful by providing a detailed description of the challenge to be solved including a prioritized set of needs. This allows the agencies to have solid control over the types of inputs they receive. For instance, in the example we used earlier, the city can put forth the challenge for citizens to submit ideas on how to alert rural citizens of an impending weather event. This is much more effective than the open-ended "how can we improve" sites that many government agencies employ as a means to get input from their citizens. Granted, that the government is asking for citizen input at all is a great step forward; it demonstrates that they now view the citizen/taxpayer as a customer who has valuable input. Now, take it a step further

---

* NPR story, October 2, 2011, "When Scientists Fail, It's Time to Call in the Gamers," NPR Staff.

and give the citizens a good target to hit, a detailed challenge, and you'll be surprised at the great ideas that arise—ideas that solve the problems *that need to be solved.*

In Summary

I realize that this chapter was a whirlwind of information—too much to cover. The topic of innovation in the government sector could be an entire book on its own. My goal is to demonstrate the need to apply standardized techniques and discipline within the government framework so that innovation programs can be as effective as possible. In our economy, we don't have a spare dollar to waste on guesswork. We need answers and solutions that work.

NOTES

Foreword

1. Peter Drucker, "The Discipline of Innovation," *Harvard Business Review*, 1985, p. 102 (as seen in the August 2002 edition of HBR).

Introduction

1. David Garvin, Harvard Business School.

2. Bill Gates, "Creative Capitalism," *Time*, July 2008.

3. James Philis et al., "Rediscovering Social Innovation," *Stanford Social Innovation Review*, Fall 2008.

4. Andrew Wolk, "The Emerging Social Impact Market: Fostering Social Innovation and Investing in What Works," speech at the 2010 Nonprofit Management Institute, Palo Alto, CA, October 5–6, 2010.

5. Michael E. Porter and Mark R. Kramer, "The Big Idea: Creating Shared Value," *Harvard Business Review* 89, no. 1–2, January–February 2011.

6. Ibid., page 4.

7. YouTube video from an interview during the Sustainable Brands 2009 conference. Quote is from 2:15 in the video http://wn.com/Green_Economic_Recovery_Andrew_Winston.

8. "President Obama to Request $50 Million to Identify and Expand Effective, Innovative Non-Profits," press release, May 9, 2010, http://www.whitehouse.gov/the_press_office/President-Obama-to-Request-50-Million-to-Identify-and-Expand-Effective-Innovative-Non-Profits/. Accessed January 5, 2011.

9. ITI Scotland.com, "About Us," http://www.itiscotland.com/defaultpage121c0.aspx?pageID=33. Accessed January 5, 2011.

10. Mary Coughlan and Jimmy Devins, *Innovation in Ireland* (Dublin: Department of Enterprise, Trade & Employment, 2008), http://www.deti.ie/publications/science/innovationpolicystatement.pdf. Accessed January 5, 2011.

Chapter 1

1. John C. Camillus, "Strategy as a Wicked Problem," *Harvard Business Review* 86, no. 5, May 2008.

2. John Kania and Mark Kramer, "Collective Impact," *Stanford Social Innovation Review*, Winter 2011, p. 36.

3. Clayton Christensen, Michael Horn, and Curtis Johnson, *Disrupting Class* (New York: McGraw-Hill, 2008), p. 11.

4. Clayton Christensen, Heiner Baumann, Rudy Ruggles, and Thomas Sadtler, "Disruptive Innovation for Social Change," *HBR*, 2006.

5. Scott Anthony, Mark Johnson, Joseph Sinfield, and Elizabeth Altman, *The Innovator's Guide to Growth*, (Boston: Harvard Business Press, 2008), p. 45.

6. C. K. Prahalad, *The Fortune at the Bottom of the Pyramid: Eradicating Poverty Through Profits* (Upper Saddle River, NJ: Prentice Hall, 2009), p. 35

7. Unite for Sight website, Module 2: Social Marketing at the Base of the Pyramid, http://www.uniteforsight.org/social-marketing/base-of-pyramid.

8. Ibid.

9. Wordpress Application Program Interface, February 13, 2011.

10. Susie Boss, "Under One Roof," *Stanford Social Innovation Review*, Spring 2011, p. 63.

11. Ibid.

12. Tandem Strategies, Inc., "The Innovation Partners." The Innovation Partners was created as a spinoff from the Strategyn, Inc., family of companies in 2010. Its purpose is to bring the concepts of outcome-driven innovation and other state-of-the-art innovation techniques to address social issues. Our discovery needs archives (DNA), data sets composed of the jobs and outcomes of the ecosystem members in several scenarios, have provided a foundation for a number of effective innovation programs nationwide. Through the development of these archives, organizations can quantify the data in their specific geographic area, thereby significantly increasing their speed of innovation. Once the jobs and outcomes relating to a given social problem have been captured, those data can be reused by others attempting to deal with the same problem in other geographic or demographic areas. Anthony Ulwick gives an in-depth explanation of this process in Anthony W. Ulwick, "Turn Customer Input into Innovation," *Harvard Business Review* 80, no. 1, January 2002, pp. 91–97. We use several techniques to ensure that respondents are answering the questions truly and not just clicking on responses. One that has worked well for us is to establish a minimum standard deviation test for both importance and satisfaction. We eliminate from the sample the surveys of respondents whose answers do not have adequate discrimination. For more detailed information on outcome-based segmentation, please see Chapter 4 of *What Customers Want*, Ulwick, 2002.

13. Anthony Ulwick, *What Customers Want* (New York: McGraw-Hill, 2002), p. 25.

14. Lance Bettencourt and Anthony W. Ulwick, "The Customer-Centered Innovation Map," *Harvard Business Review* 86, no. 5, May 2008, pp. 109–114.

15. Anthony Ulwick gives an in-depth explanation of this process in Anthony W. Ulwick, "Turn Customer Input into Innovation," *Harvard Business Review* 80, no. 1, January 2002, pp. 91–97.

16. There is a time in which the outcomes are the appropriate level of detail and should be used—situations in which the market is very mature and there is already substantial satisfaction for the current products and services. In such situations, an organization is looking for any hidden opportunity that is often found at this granular level. An example of this is the Bosch circular saw example where the market was very mature, and products have been around for decades. In this case, the team had to really dig into the use of the saw to identify areas of weakness. The outcomes are extremely effective at identifying these kinds of niche opportunities. However, in most social scenarios, this is not the case. Generally the issues are so substantive that there is no problem finding ample opportunity within the job at the job step level.

17. For more detailed information on the structure of job statements, please refer to Ulwick, *What Customers Want*, Chapter 2.

Chapter 2

1. For more detailed information on capturing customer needs, see Chapter 2 of Ulwick, *What Customers Want*.

2. Tim Brown and Jocelyn Wyatt, "Design Thinking for Social Innovation," *Stanford Social Innovation Review*, Winter 2010, pp. 31–33.

3. Lance Bettencourt and Anthony Ulwick, "The Customer-Centered Innovation Map," *Harvard Business Review* 86, May 2008, p. 109.

4. Anthony Ulwick, *What Customers Want* (New York: McGraw-Hill, 2002), pp. 15–38.

5. For more detail on the outcome gathering process, please refer to Ulwick, *What Customers Want*, Chapter 2.

6. The method outlined by Ulwick in *What Customers Want* recommends that all outcome statements be rated for importance and satisfaction. Because there are typically between 100 and 150 outcome statements, it becomes a very difficult and tedious questionnaire to complete, not to mention expensive. During my tenure with Strategyn and executing this methodology, we had a 100 percent success rate of having the customers fill out the questionnaires. I believe that there is a less burdensome option that yields virtually the same level of precision. The simplified version of the needs statements saves time for the respondent to fill out the questionnaire, results in a less expensive questionnaire, reduces respondent fatigue, and reduces the dropout rate (where respondents start the questionnaire but then fail to complete it). This simplified version is especially vital when dealing with social issues since, in these cases, we are dealing with at least two (often

more) audiences, and in many cases inexpensive Web-based studies are not an option for the audience being studied.

Table N-1 shows the differences between the standard outcome-driven innovation (ODI) approach and the social innovation quantitative approach.

Ta b l e   N - 1 .   Quantitative Approach Comparison

Scope of Initiative	Recommended Approach for Quantitative Survey			
	Process	Advantages	Disadvantages	When to Use
Outcome-Driven Innovation	Job steps are included in the questionnaire for headings of the outcome statements to be rated. The job steps themselves are often not rated, but each outcome statement is rated for importance and satisfaction. Contextual interviews are used to follow up in order to gain better understanding of the outcomes that rated as a high opportunity.	Very thorough results as to specifically which outcomes are problematic for the audience.	Extremely lengthy survey, respondent fatigue, cost of survey.	When dealing with a job that you are very unfamiliar with and when working in an area of high commoditization where you are looking for hidden opportunities.
Social Innovation Method	Rate the job steps for importance, satisfaction, and the constraints. Use contextual interviews to dive deeper into the job step to identify whether the step is unsatisfied due to speed, complexity, error, etc. Therefore, in this case the outcomes are gathered in a qualitative format after identifying the areas of the job that are important and unsatisfied.	Significantly reduced questionnaire length and cost. Allows more questions to be asked about the steps of the jobs, related jobs, emotional jobs, and constraints. The questionnaire can also be used to gain perspective on the baseline measure of the jobs—what the consumers' expectation is for how long it should take to execute a job step, whether they will pay for that job to be improved, etc.	Outcomes are not quantified and therefore not prioritized. The trade-off here is whether this level of detail is needed. If it is, then we recommend proceeding with the traditional ODI method.	When investigating social scenarios at their big-picture ecosystem level.

7. We use several techniques to ensure that respondents are answering the questions truly and not just clicking on responses. One that has worked well for us is to establish a minimum standard deviation test for both importance and satisfaction. We eliminate from the sample the surveys of respondents whose answers do not have adequate discrimination. The second test we use is the time it takes to complete the survey. We have our vendors take the survey—just reading it and not necessarily answering the questions. This is the fastest time allowed as it indicates how fast the survey can actually be read, much less answered. If the respondent completes the survey in less time than this benchmark, the respondent is eliminated from the study. We generally see about a 10–15 percent elimination rate which needs to be accounted for in the sample design. So, if your goal is 600 completes, you would want to ensure that you have 660–690 surveys to allow for the respondents who will be eliminated.

8. This is different from the technique proposed in *What Customers Want* which aggregates the opportunity across the sample, or a cut of the sample. For instance, in the method proposed by Ulwick in *What Customers Want*, the importance scores and satisfaction scores represent the percentage of the sample that scored an item in one of the top two boxes, *each calculated independently*. Therefore, if 75 percent of the people rated an item top two box importance and 35 percent rated it satisfied, there is no way to know how much of the 75 percent of the people who rated it important *also* rated it unsatisfied, since these calculations are conducted independent of each other. Of the 65 percent who rated the item *less* than top two box satisfied, some of them may be among the 75 percent who found the item to be important and some may not be. In Ulwick's model, the opportunity algorithm is then applied to these percentages. In contrast, the individual opportunity score itself is illustrated as a percent of the sample that rated the item *both* important *and* unsatisfied, ensuring that we are looking at the actual opportunity potential, the marginal opportunity potential, and the strength of the existing market. Since we also explore the entire spectrum of how people responded to the need statement, we have even more information to determine to what degree that statement is one that should be pursued for value creation across the ecosystem.

Chapter 3

1. Ellen Meara and Merideth Rosenthal, "Comparing the Effects of Health Insurance Reform Proposals: Employer Mandates, Medicaid Expansion and Tax Credits," *The Berkeley Electronic Press-Forum for Health Economics and Policy*, February 2007.

2. AAFP website information, http://www.aafp.org/online/en/home/practicemgt/quality/qitools/pracredesign/january05.html.

3. Arif Ahmed and Jack E. Fincham, "Physician Office vs. Retail Clinic: Patient Preferences in Care Seeking for Minor Illnesses," *Annals of Family Medicine* 8, 2010, pp. 117–N123.

4. National Public Radio transcript, June 27, 2011, "The Parkinson's Doctor Will Video Chat with You Now," http://www.npr.org/2011/06/27/137089619/the-parkinsons-doctor-will-video-chat-with-you-now.

5. Jean-Herve Bradol and Claudine Vidal, *Medical Innovations in Humanitarian Situations*, Doctors Without Borders/Médecins Sans Frontières (MSF) 2011, pp. 3–17.

Chapter 4

1. Sarah Caldicott-Miller, "Ideas-First or Needs-First: What Would Edison Say?" whitepaper, http://www.strategyn.com/resources/white-papers/, 2009.
2. "Investing in Social Growth," whitepaper produced by the Young Foundation, September 2010.
3. Ellen Domb, "Titanic TRIZ: A Universal Case Study," published in the proceedings of the 2nd Altshuller Institute Conference, May 2000.
4. Edward de Bono, *Lateral Thinking: Creativity Step by Step*, (New York, NY: Harper & Row, 1970), pp. 9–13.
5. Kalevi Rantanen and Ellen Domb, *Simplified TRIZ* (Boca Raton, FL: CRC Press, 2002), p. xvi.
6. While The Next Idea is currently inactive because the principals have left to pursue other careers, the tools I show were originated with that organization under either Terry Richey's (former CEO and Founder of The Next Idea) leadership and design or my own. Therefore, I still refer to them as "The Next Idea" tools.
7. Tim Brown and Jocelyn Wyatt, "Design Thinking for Social Innovation," *Stanford Social Innovation Review*, Winter 2010, pp. 31–33.
8. Satish Nambisan, "Platforms for Collaboration," *Stanford Social Innovation Review*, Summer 2009, pp. 44–47.
9. John Kania and Mark Kramer, "Collective Impact," *SSIR*, Winter 2011, p. 36.
10. Satish Nambisan, "Platforms for Collaboration."
11. Ibid.
12. Ibid.
13. Personal interview with Brandon Knicely, partner in the complex at Austin Innovation Partners, July 2011.

Chapter 5

1. Bill Gates, "Making Capitalism More Creative," *Time*, July 31, 2008.
2. Michael E. Porter and Mark R. Kramer, "The Big Idea: Creating Shared Value," *Harvard Business Review* 89, no. 1–2, January–February 2011.
3. Ibid.
4. Alexander Osterwalder and Yves Pigneur, *Business Model Generation* (Hoboken, NJ: John Wiley & Sons, 2010), pp. 15–20.
5. John Kania and Mark Kramer, "Collective Impact," *SSIR*, Winter 2011, p. 36.
6. John Kania and Mark Kramer, "Catalytic Philanthropy," *SSIR*, Fall 2009, p. 30.
7. "President Obama to Request $50 Million to Identify and Expand Effective, Innovative Non-Profits," press release, May 5, 2009.

8. Website, Massachusetts Gov. Administration and Finance, "MA Pursues Social Innovation Financing to Spur Innovation and Build on Program Success" May 6, 2011.

9. Ibid.

10. The Nature Conservancy website http://www.nature.org/ourinitiatives/regions/southamerica/colombia/howwework/water-fund-bogota.xml.

11. Ibid.

12. Ibid.

13. Michael E. Porter and Mark R. Kramer, "The Big Idea: Creating Shared Value," *Harvard Business Review* 89, no. 1–2, January–February 2011, p. 2.

Chapter 6

1. Susan Evans and Peter Clarke, "Disseminating Orphan Innovations," *Stanford Social Innovation Review*, Winter 2011, p. 43.

2. José Guimón and Pablo Guimón, "Innovation to Fight Hunger: The Case of Plumpy'Nut," Accenture working paper no. 2010/01.

3. Susan Evans and Peter Clarke, "Disseminating Orphan Innovations," *Stanford Social Innovation Review*, Winter 2011, p. 43.

4. Ibid.

5. Ibid.

6. Jeffrey Bradach, "Scaling Impact," *SSIR*, Summer 2010, pp. 27–29.

7. National Institute for Technology in Liberal Education, Web site, URL: http://www.nitle.org/help/

8. Bradach, "Scaling Impact," *SSIR*.

Chapter 7

1. Clayton Christensen, Jerome Grossman, and Jason Hwang, *The Innovator's Prescription* (New York: McGraw-Hill, 2009), p. xxiii.

2. Ibid, p. 163.

3. This list of constraints is a small subset of the total list obtained by physicians in a study conducted in 2010 in the eastern portion of the United States.

4. Christensen, Grossman, and Hwang, *The Innovator's Prescription*, (New York: McGraw-Hill, 2009) p. 122.

5. Ibid, p. xxiii.

6. Seth Godin, *Poke the Box*, (The Domino Project, 2011).

Chapter 8

1. Peter Kareiva, "The Future of Conservation: Balancing the Needs of People and Nature," *The Nature Conservancy*, Spring 2011, p. 32.

2. Andrew Winston, "Is Water the Next Carbon?" *Harvard Business Review* online blog: http://blogs.hbr.org/winston/2011/01/is-water-the-next-carbon.html.

3. Andrew Winston, "Making Green Work," *Harvard Business Review,* February 9, 2011, pp. 3–4.

4. Ibid, p. 3.

5. Bob Pennisi and I conducted this study when we were working at Strategyn, Inc.

6. Clayton Christensen et al., "Picking Green Tech's Winners and Losers," *Stanford Social Innovation Review,* Spring 2011, pp. 30–35.

7. Ibid., p. 35.

8. This figure is based on two to three minutes of brushing, rinsing, and so on while leaving the water running. Fact obtained from http://gogreeninyourhome.com/water-conservation/how-to-save-water-brushing-teeth/.

9. Christensen, "Picking Green Tech's Winners and Losers," *Stanford Social Innovation Review,* p. 35.

10. The Street website, http://www.thestreet.com/story/11200247/1/building-the-infrastructure-for-widespread-renewable-energy.html.

11. John Shimkus, "Building the Infrastructure for Widespread Renewable Energy," *Energy Digital,* July 27, 2011.

12. Peter Kareiva, "The Future of Conservation: Balancing the Needs of People and Nature," *The Nature Conservancy,* p. 39.

13. Ibid., p. 40.

14. Ibid.

15. The Nature Conservancy website http://www.nature.org/ourinitiatives/regions/southamerica/colombia/howwework/water-fund-bogota.xml.

16. Ibid.

17. Ibid.

18. Going Green, study, January 2008, Strategyn, Inc., in conjunction with CABA.

Chapter 9

1. Satish Nambisan, "Transforming Government through Collaborative Innovation," published by the IBM Center for The Business of Government, 2008, p. 6.

2. John Kania and Mark Kramer, "Collective Impact," *Stanford Social Innovation Review,* Winter 2011, p. 36.

3. Ibid, pp. 39–40.

4. Joby Warrick, "Little Blue Pills among the Ways the CIA Wins Friends in Afghanistan," WashingtonPost.com, December 26, 2008.

5. Alyssa Battistoni, "An Ounce of Advocacy," *Stanford Social Innovation Review,* Winter 2010, pp. 37–39.

6. "Barack Obama Candidate Platform," *Boston Herald,* December 27, 2007, http://www.bostonherald.com/news/national/politics/2008/bios/view.bg?articleid=1063110.

INDEX

ABOUT THE AUTHOR

Sandy Bates has spent the last decade as a strategy consultant specializing in helping clients realize the value of innovation as a means to grow existing businesses, reposition current offerings and identify new market spaces. In her work with dozens of industries, and more than 100 innovation initiatives, she has explored, developed and tested innovation methods and theories in real world situations. As such, she has helped executive teams launch award-winning products, services, and programs. Sandy most recently founded The Innovation Partners where she began applying the leading edge methodologies of Strategyn and others to the social issues plaguing our society and our world, including challenges such as health care for the uninsured, natural resource conservation, education, care for the elderly, and many others.

Before the development of The Innovation Partners, Sandy was the director and cofounder of the Strategyn Institute where she trained and mentored hundreds of executives in the Outcome-Driven Innovation methodology. During her tenure with Strategyn Sandy also served as the lead consultant for Microsoft, conducting over 30 innovation initiatives in the consumer, software development, and business applications divisions as well as exploring new market areas. She has worked extensively in the health care field with companies such as Guidant, Medtronic, WellPoint, Hospira, Lancaster General Health, the California Healthcare Foundation and many others. Sandy also worked with the Harvard/Kennedy School of Government's study on health care delivery models in America. Other clients have included Morgan Stanley, Masco, Dell, AIG, State Farm, Hill's Pet Nutrition, Cargill, J. R. Simplot Company, and Hallmark. Sandy specializes in ideation, services innovation and messaging / positioning.

Prior to joining Strategyn, Sandy was a principal of The Next Idea, a marketing innovation consulting firm. Her past experience in corporate marketing focused on strategy, market positioning, product development, services marketing, and primary research. She served as chief marketing officer in the technology services and health care industries. Her experience includes the successful execution of an IPO, redefining a market space for a mid-size enterprise, and sustaining growth and brand presence through a major merger at a Fortune 500 company.

She graduated summa cum laude from Arizona State University with a bachelor of science in psychology and has pursued postgraduate studies in services marketing, competitive strategies, direct marketing strategies, and product innovation. She calls Austin, Texas her home base where she lives with her husband Rob and enjoys the blessings of raising three young men.